Endorsements for *Lasting Change*

This is a timely and substantive work which can profoundly influence your business and enhance your life.

Nido R. Qubein, Chairman, Creative Services, Inc.

This is the essential traveler's guide for companies embarking on a journey of Shared Values. Rob Lebow and Bill Simon mentor unselfishly in this rich resource of a book. Leaders who are serious about changing the context of their organizations to create the Heroic Environment will start here.

Joel Marks, corVISION MEDIA

If your purpose in life includes creating a workplace for people who want to love their work, then, at this point in time, Rob Lebow has "The Answer"!

Maurice Orr, President, Orr Automotive

As we approach a new milennium, values driven companies are a mandate for a healthy world view and economy! *Lasting Change* is a strong declaration and a source of implementation for companies of all sizes!

Naomi Rhode, VP, SmartPractice, past President, NSA

I find the new publication, *Lasting Change,* to be an interesting and honest venture on the road to the heroic environment needed in today's business world.

James L. Martineau, Executive VP, Apogee Enterprises

This book is a credit to the lasting power of people and business values. The eight principles are the foundation for our business and the starting point for our issues management consulting business.

John Eastham, EMB Partners

Lasting Change

The Shared Values Process™

That Makes Companies Great

Lasting Change

The Shared Values Process™

That Makes Companies Great

Rob Lebow
and William L. Simon

VAN NOSTRAND REINHOLD
IⓣP® A Division of International Thomson Publishing Inc.

New York • Albany • Bonn • Boston • Detroit • London • Madrid • Melbourne
Mexico City • Paris • San Francisco • Singapore • Tokyo • Toronto

Printed in the United States of America

www.vnr.com

For more information contact:

Van Nostrand Reinhold
115 Fifth Avenue
New York, NY 10003

Chapman & Hall GmbH
Pappalallee 3
69469 Weinham
Germany

Chapman & Hall
2-6 Boundary Row
London SEI 8HN
United Kingdom

International Thomson Publishing Asia
60 Albert Street #15-01
Albert Complex
Singapore 189969

Thomas Nelson Australia
102 Dodds Street
South Melbourne 3205
Victoria, Australia

International Thomson Publishing Japan
Hirakawa-cho Kyowa Building, 3F
2-2-1 Hirakawa-cho, Chiyoda-ku
Tokyo 102 Japan

Nelson Canada
1120 Birchmount Road
Scarborough, Ontario
M1K 5G4, Canada

International Thomson Editores
Seneca, 53
Colonia Polanco
11560 Mexico D.F. Mexico

1 2 3 4 5 6 7 8 9 10 QEBFF 02 01 00 99 98 97

Library of Congress Cataloging-in-Publication Data available

ISBN 0-442-02585-8

Those who talk only of people are small-minded;
    shun them.
Those who talk of events are thinkers;
    listen to them.
Those who talk of ideas are leaders;
    follow them.

William L. Simon

To Arynne, Victoria, and Sheldon
and to Sharon and Lauren

Acknowledgments

From Bill Simon

It seems highly appropriate to begin this book on values by acknowledging my wife, Arynne, whose exceptional set of values includes fun and adventure along with kindness, caring, and the responsibility to add beauty and quality to the world. She is an inspiration not just to me and to our children, Victoria and Sheldon, but to the many people she teaches as a mentor, consultant, and writer. My precious wife is a super model of values to live by, a fact well recognized by the many people whose lives are the better because of her.

My gratitude to our editor, John Boyd—now Editorial Director at VNR—who was strongly committed to this book from the very first.

There are other people who deserve my recognition and appreciation for their help with extensive research. Jennifer Larson is the admin who supports both Arynne and me; in addition to juggling our typically pressured schedules, she manages to keep any number of librarians busy tracking down the facts and data, and then she somehow manages to keep these growing stacks of research documents organized. On a daily basis, in a gracious and kind way, she protects me from interruptive phone calls without injuring old friendships.

Upstream of the copy editing by New Yorker Shelley Flannery, a sharp-eyed West Coast editor named Jane Alberts proofed the manuscript; both offered more valuable suggestions than I had any right to expect. A good copy editor is hard to find, and the manuscript flows just a little better and reads just a little more smoothly thanks to their suggestions.

And then there is John Kistner, of Rob Lebow's staff, who volunteered his services as a beyond-the-call-of-duty effort. John's style is impressively tenacious and I'll never know how he manages to locate obscure pieces of accurate information in such short order.

Finally, my thanks to co-author Rob Lebow for keeping after me and keeping his enthusiasm high until we found a way to do this book together.

From Rob Lebow

In Memory of Ruth Lebow—who would not have been surprised that her son has so much to say and finds a way to say it.

In 1918, my father left school to help support his family. His job was carrying a pail of hot coals to keep rivets at the right temperature for the riveters in the Boston Shipyards. He recalls how bitter the winds and the work felt to him at age 15. But now, at age 93, when Dad talks about those days, you can see a sparkle of pride in his eyes as he recalls building destroyers to help support the war effort.

His work in the shipyards didn't last long. For the past 70 years, my father's contribution has been through his artistic talents. In the 1960s, his ads for *Newsweek* could be seen every Monday in major newspapers around America. Maybe you or your parents responded to the public service ads he created for U.S. Savings Bonds and other national causes, which appeared in newspapers and on rapid transit vehicles. Dad prides himself on their artistry as well as their effectiveness.

He still paints vigorously and lives on his own, while lighting up the lives of those around him with his boundless creativity. If you could meet my Dad, his unassuming manner would not expose the real story of who he is and he wouldn't tell you about the thousands of people he has touched and helped. His name has never been up in lights but he has electrified my life. Dad was my first hero and has always been my primary role model—kind, caring, and unconditionally there for me.

The other hero in my life is my wife, Sharon. Why she has put up with my unique ways of thinking all these nearly 30 years, I shall never know. But she's stuck by me through thick and thin and is, as many other lucky men would say, my most honest critic. For finding her I shall always be grateful. And to my daughter, Lauren, who at age 13 has developed the ability to accept the reality that to see me usually requires a trip to the office, I am grateful for her love and understanding.

One does not just sit down and write a book. I found it even more demanding than my co-author, Bill Simon, warned. His patience and laid-back style were a perfect balance to my high intensity; we have not only written a book but built a fine relationship. Arynne Simon, Bill's wife, is one of those people whose every thought and idea is worth writing down; I continue to delight in her wisdom, humor, generosity, and forthrightness.

John Boyd at VNR is a rare editor, indeed—inspiring, uplifting, and supportive. John Kistner and editor Jane Alberts, as well others on our Seattle staff, helped in various ways to improve the manuscript.

As for the many clients and consulting distributors we have come to know over the past ten years, we are grateful to them for having the patience to listen while the Shared Values ideas were still being shaped. I know these early adopters will be proud of this book because they took a risk and courageously pioneered the Shared Values Process with us.

I dream of a future that includes quality values woven into the tapestry of life and I am grateful to all of you who share my dreams.

We were born to manifest the glory of God
　　that is within us.
It is not just in some of us; it's in everyone.
And as we let our own light shine, we unconsciously give
　　other people permission to be the same.
As we are liberated from our own fears, our presence
　　automatically liberates others.

Anonymous

Contents

FRAMING YOUR SYSTEMS TO PEOPLE NEEDS AND CUSTOMER NEEDS • *191*

THE PATH TO LASTING CHANGE • *241*

Values Under Fire:
The Pepsi Case

A crisis will quickly test the strength and fiber of any organization. About lunchtime on Thursday, June 10, 1993, in the midst of its peak selling season, the Pepsi-Cola bottler in Seattle received a jarring phone call that would echo across America and explode into an international news event.

Taking the call was Carl Behnke, CEO of Alpac Corporation. On the other end of the line was a newsman asking for a statement. Behnke was alarmed to learn that a distraught 82-year-old man said he had found a hypodermic syringe in a can of Diet Pepsi. As Pepsi's regional bottler for the Northwest, Alpac was in the direct line of fire.

Yet Behnke knew it would be virtually impossible for a foreign object to get into a can of Pepsi at the plant; even a disgruntled employee would be hard put to defeat the processes and protections.

You know the claim is a phony, you know an investigation will prove that the person who says he found the object is curiously mistaken, seriously disturbed, or committing blatant fraud. So what do you do? Stonewall? Put out a first-strike press release arguing the claim is impossible?

A descendant of the Skinners, who were, like the Nordstroms, Weyerhausers, and Boeings, one of the original Seattle families, Behnke lived with a deep commitment to his community. And he remembered all too well how ineptly Exxon had handled the *Exxon Valdez* oil spill just four years earlier—so badly that public antagonism had forced the company to sell its retail gas stations in the Pacific Northwest.

Now it was Behnke's turn to react; without warning, he had become a celebrity, and it looked as if this unwelcome spot in the limelight was going to last a good deal longer than Andy Warhol's promised 15 minutes.

In the years since taking over as president of Alpac, Behnke had sought to build a company with a social conscience. One morning in early 1990, after he had learned of the work of our consulting group, he had challenged his senior team to join in taking the company on a journey toward a "Heroic Environment." The cost of the effort was projected to be two cents per case of soda, and he asked, "Will we get two-cents-a-case worth of value out of investing in a process based on Shared Values?" Saying "No" to Carl Behnke once he had decided on a course of action was always hard—his personal charisma went far in winning people to his view. That morning he had put the question in financial terms; at a cost of only two cents a case, no one could turn the proposition down. We soon began the process of bringing all Alpac employees to the practice of Shared Values.

Forty-two months later, when the crisis hit, the executives got an unmistakable confirmation that they had made the right choice in beginning that journey.

Two of the eight basic Shared Values defined the ground rules here: *Put the interests of others first,* and *Treat others with uncompromising truth.* To CEO Behnke this meant, "We had to operate under the premise that the tampering could have happened in our plant until we could prove differently." The company immediately notified the Food and Drug Administration, but Behnke recalls the FDA people "were perturbed that the consumer's lawyer had told the story to the press," fearing the report might trigger copycat complaints.

The next day around 3 P.M., another customer reported finding a syringe, this time in a can of Mountain Dew that had been produced months earlier—making it almost certain this was exactly what the FDA had feared, a copycat claim. The individual approached Alpac directly, suggesting the contact be kept quiet, but Alpac recognized its obligation to alert everyone immediately: the crisis would not be handled behind closed doors. Together the company and the FDA prepared a joint advisory recommending that consumers pour their Diet Pepsi into a glass before drinking it; Behnke asked to take the lead role in informing consumers, and the agency agreed. That evening Behnke went on TV, live.

Why would Alpac executives alarm their own customers like this? Simple: they were putting consumer safety and the interests of others first, as their corporate shared values required— despite knowing there was only one chance in a million, a hundred million, that the reports could be true.

With the second report, the FDA sent in a team that embarked on a thorough examination of the plant, production records, and quality-assurance processes. Meanwhile a flood of press calls began to pour in. Alpac fearlessly threw its doors open, inviting reporters and TV crews to tour the plant and see for themselves the process in which each can is turned upside down, blasted clean with a powerful jet of air or water, turned back up, filled, and closed—all within nine-tenths of a second.

The Pepsi scare quickly hit the headlines coast to coast as 60 people all over the United States filed reports of syringes in their soft-drink cans. Over the next several days, sensationalist coverage led the network evening news programs. Nationally, Pepsi-Cola lost $25 million in the first week alone.

But the tone of the stories changed just as quickly, as the authorities, including FDA commissioner David Kessler, began making comments favorable to Pepsi's quality assurance and commitment to protect consumers. The national media, not usually known for a pro-business stance, weighed the evidence and began asking questions about the validity of the consumers' claims. Soon newscasters were reporting arrests; in time, some 60 reports would prove to be fraudulent.

What caused the press to turn around so quickly and become advocates for Pepsi-Cola? Pepsi had handled the situation heroically, in an open, forthright manner, valuing the public interest and safety above company prestige and financial interests. And public support rebounded quickly: within a week, sales were up by 800,000 cases nationally; Pepsi sales for the Fourth of July week set a record for the year, while PepsiCo stock went up more than a point. Carl Behnke and his organization were seen as honorably responsive.

The PepsiCo attitude was summed up by corporate CEO Craig Weatherup: "Public safety was paramount to all other concerns. Our strongest allies during the crisis were honesty and openness." Values, not tactics, had ruled the actions of the company and its people. In the end, what mattered most was an ingrained organizational commitment to Shared Values.

Advertising Age called the incident "a textbook case of how to come through a PR crisis," and the company was even lauded by the FDA for its handling of the situation. Pepsi-Cola corporate received accolades on the floor of the U.S. Congress as well as a national consumer award for an excellent example of corporate citizenship.

PepsiCo had lived up to its vision statement, which includes the words "We will be . . . guided by Shared Values." An outside observer, not knowing about Shared Values and the power values hold, could have believed in the early hours that the crisis could go either way. Indeed, it could have been another *Exxon Valdez*.

John Eastham, then president of Alpac's outside public relations firm, vividly remembers those terribly stressful days and has his own answer for the reason it turned out as well as it did: "Like Alpac, we had brought the Shared Values philosophy into our firm and it gave us the guidance. For Carl Behnke and me, there was only one answer: telling the truth. When values lead your actions, decisions are easy."

Preface

In a time frame just short of a decade, tens of thousands of people around the world have accepted that working in a Heroic Environment can boost productivity and profitability while creating a workplace with positive feelings about individual contributions and the ways that individual employees relate to others. These people now work in energized surroundings where positive values are widespread and where successes are enthusiastically shared by all.

We call this a Heroic Environment because the idea of putting the interests of others first is a concept virtually unique in the workplace. It is a Heroic Environment because self-expression, a can-do attitude, and a passion for playing at the top of your game are the expected behaviors instead of a surprising exception.

WHY COMPANIES FAIL TO PROSPER

A wise man suggested that the best working definition for insanity was "continuing to do the same thing you have always done while expecting a different outcome."

Not seeking new and improved ways to handle your organizational challenges puts you in a high-risk situation that could ultimately force you out of business. Even for a publicly funded or nonprofit organization, once you lose the ability to serve the customer, the marketplace will turn away from you.

Think that sounds overly dramatic? Take a look at the statistics. *Fortune* magazine has been keeping records on the top

500 U.S. companies since 1950. In that half-century, only 11 percent of the original 500 companies still exist under their own banner. The others have been driven out of business or have merged with a stronger, more dominant player.

And what was the one common denominator that destabilized these once-outstanding organizations? In almost every case, it was an internal malaise that quietly undermined the vision, flexibility, and culture of the company. A *people* problem was the iceberg for these *Titanics*.

Ten years of watching hundreds and hundreds of organizations, large and small, try to remake themselves should have sent a clear message to business leaders who have been putting their trust in programs like Total Quality Management, reengineering, rightsizing, restructuring, and culture change. If you've been in the business world for any period of time, you've seen your own organization buy into an assortment of programs and interventions aimed at improving morale, enthusiasm, teamwork, sales, quality, creativity, and conflict resolution. Each was hailed as the silver bullet to slay the demons keeping your operation from achieving the success that seemed just beyond grasp.

Repeatedly the message went out, "If only we all embrace this new program, everything will be improved." And each time another new banner was displayed in the lobby, the implicit promise was that by acquiring the right set of work habits, by adopting the right approach to knowledge, by correcting everyone's thinking, emulating a ninja-warrior philosophy, or introducing the right reengineering strategy, the people of your company would be saved and the business would prosper beyond imagination. But it was the corporate version of the cult mentality, a setup for repeated failures.

Many of these programs did bring results, albeit modest ones. But almost always they have left the companies still wanting a cure for a range of remaining distresses. Unless the changes were begun with a focus on people at every level, the efforts to remake or transform the business in the hopes of competing more effectively have met with disappointing results.

THE EIGHT HEROIC VALUES

More than ten years ago, I began leading a team on a quest to find the reason that organizations were not reaping the reward of established training. We set out to answer a simple but challenging question:

What do people want in their work environment to be more productive and to perform at the top of their game?

Our team spoke to and interviewed dozens of business people, HR professionals, instructional designers, business consultants, training companies, researchers, business leaders, CEOs, organizational development specialists, and business journalists. We scanned the libraries for every scrap of pertinent data and haunted bookstores for new thoughts and ideas about what people wanted in order to become more productive. After all the searching, we knew that no person or organization had a satisfactory answer.

Finally we stumbled upon a research project that had begun at the University of Chicago, in which the team had collected 17 million surveys of workers in 40 countries around the world on what people wanted in their work environments to be productive, creative, fulfilled, and competitive. The data outlined what it would take for everyone in an organization, regardless of the job or industry, to play at the top of his or her game.

We asked, "What are the components of the work environment that people want?" The answers were right there in the database: eight items that we have since come to call the eight Shared Values. Yet we almost missed the significance; the list seemed too easy, too obvious—almost like sitting down with a zealous Bible student and asking, "What are the virtues the Bible teaches us to practice?"

Luckily we realized the significance: these eight Shared Values revealed what 17 million workers had said they wanted in order to feel fulfilled in the workplace. And our nearly 100,000 surveys since then, with employees, managers, and corporate executives, confirm those original findings. The eight values come not from the musings of some management guru, but from the people in the workplace themselves. We eventually came to realize that if the list seems obvious, the reason is entirely valid: what employees, managers, and executives sense they are missing grows out of fundamental human needs that are virtually universal.

The Eight Shared Values of the Heroic Environment

Truth:	Treat others with uncompromising truth
Trust:	Lavish trust on your associates
Mentoring:	Mentor unselfishly
Openness:	Be receptive to new ideas
Risk-taking:	Take personal risks for the good of the organization
Giving credit:	Give credit where it's due
Honesty:	Be honest in all dealings; do not touch dishonest dollars
Caring:	Put the interests of others before your own

THE SHARED VALUES ENVIRONMENT

When the process of instilling Shared Values is launched in an organization, at first almost everyone assumes it will be just another flavor-of-the-month management program of short life span. They soon discover it is not a program, and is not short-lived.

The carefully designed process of sharing values allows people for the first time in their work experience to purposefully and carefully challenge the *context* of their work environment instead of placing blame, pointing fingers, or trying to "fix" peo-

ple as they have always done in the past. When you systematically examine the subject of Shared Values within the context
of what you do, and invite everyone regardless of rank to do the
same, the discussion changes dramatically. Individuals who
have regularly sat on the sidelines become fully engaged; people who have remained silent are heard from; and leaders and
managers discover that their people have capabilities never before recognized or imagined.

With Shared Values, the transformed behavior results from
how people at all levels take charge of problems. People come to
realize that *they* are the instruments of change, that if they wish
others to behave in a certain way, they must first model the desired behavior themselves. This one aspect of Shared Values
does more to change the work environment than anything else.

With Shared Values, constructive conversations replace old
political one-upsmanship patterns and manipulative promises.
Behaviors of recognition and respect couple with responsible
inquiries into root causes, as people seek to proactively eliminate problems.

The values of an organization are in essence not what we say
they are. The values of an organization are how people behave
every day. Behaviors in a Shared Values environment are
judged by fellow employees, management, and the company's
customers.

Experiences and stories from the leaders, managers, and
people in organizations that have adopted Shared Values suggest some important lessons. They report how Shared Values
has changed for the better just about everything in the way
they conduct their lives and their businesses. They describe
feeling more in control of their interpersonal relationships,
goals, objectives, and organizational strategy, and report with
some pride that customers and family members notice the difference. And they report that people within their organization
who had been unreachable are now enjoying uncommon levels
of involvement and even coming to work with enthusiasm.

Ten years ago, I could only voice the question and imagine an environment like this. Today, over 1,200 organizational work groups worldwide are on their journey toward such a place—the Shared Values environment you will find described in the pages of this book.

Ultimately, it's all the people in an organization who are its underlying strength and underlying weakness. How people deal with each other; the agreements they make and keep, both implicitly and explicitly; the standards that are established and supported; the values shared: these are the elements at the root of human survival, the elements that hold the key to true success and to *lasting change.*

The reward is a work environment where everyone puts the interests of others first, where managers and employees freely mentor one another, where everyone is open to new ideas, truth is common, and trust abounds.

SETTING OUT ON THE JOURNEY

1

An Operating System for People

Only through a radical shift in our thinking can we succeed in this new era. It calls for nothing but a complete break with tradition-bound ways . . . of the industrial age.

John Sculley,
former CEO, Apple Computer, Inc.

THE "I-5 CORRIDOR" PROJECT

In early 1989, a small restaurant chain with eight outlets along the I-5 Interstate corridor in Washington State faced a dilemma common to many successful small businesses. Mitzel American Kitchens had developed a strategy that worked, but it wasn't able to consistently produce and deliver its product—the meal—exactly the same way every time. At some of the restaurants, the differences in quality and portion size varied by the hour. The situation was driving everybody wild, especially owner John Mitzel, who knew more than anyone that if the chain was to expand successfully, it had to control its consistency and quality. Management had tried everything from standards manuals to incentives to threats, but the inconsistencies remained. Mitzel described his dilemma:

> We knew a large proportion of our business was repeat customers. Our staff are friendly people, and they take their job seriously. But the capstone to our formula is our pies—the customers love our pies.
>
> The chain was successful; I wanted to expand. But we couldn't get a hamburger stacked the same way in every restaurant, we couldn't get pie cut the same way, and these differences were blocking us from moving ahead.
>
> Cutting a pie would appear to be simple, right? It's not. I received a complaint letter from a customer who was really sore about how he had been treated. He and his wife were regular customers at our restaurant in Bellingham, where they always got a super big piece of pie. They ate one meal in our Tacoma restaurant, and wrote to let me know personally what a disappointment it was: when the pie was served, it was an ordinary-sized slice instead of the whopping piece they were accustomed to. Just one complaint letter about just one piece of pie, but the same story was being repeated over and over, and we just didn't seem to be able to control it.
>
> Some of our staff were cutting big pieces of pie to win their regular customers' hearts and keep them coming back. But it didn't net out for the company. We were wasting a lot of product because of this unwritten special treatment, and then angering customers when they only got *standard* treatment somewhere else in our chain. We were not going to expand the operation until I was confident that every staff member had a clear picture of what we had to do to be consistent and professional. If we were to expand before we shared the same values, it would only have magnified the problem. Until we could stack a hamburger the same and serve a piece of pie the same, we would have been nuts to expand.

John Mitzel is not alone in his challenge. Every business person we know does daily battle with his own hamburger-stacking and pie-cutting gremlins. The many vagaries of employees attempting new ways to serve and impress the customer cost organizations dearly. Inconsistent images of what constitutes a finished, quality job that will satisfy the customer produce a wide range of uneven, inharmonious, disruptive conditions; these well-intentioned activities occur daily in the workplace, plaguing businesses, nonprofit institutions, and government organizations without exception.

Like other organizations in the competitive restaurant industry, Mitzel's front line staff included many people who wanted to do the best thing by the customer. Yet the CEO was experienced enough to know that a beefed-up policy manual and a new set of rules weren't going to help. He had tried all the obvious control techniques and all the process improvement approaches he knew of and that several consultants had recommended—but still he was no closer to his goal of sustaining and enhancing those aspects of the business that had made the company successful when it was small, while identifying and casting off those practices that no longer worked and were getting in the way of expansion.

In many ways Mitzel was more than a typical entrepreneur. He was an explorer open to new ideas. His innovations included introducing cappuccino to his customers years before any but Italian restaurants had it on the menu. He experimented with handheld electronic meal-ordering devices before they had been proven. And he was always looking for new approaches to increase the effectiveness of the company's people, including himself.

In 1989, looking for a fresh approach to pulling everyone in his operation together, he heard about the Shared Values approach and decided he was willing to bring us in and try this new idea. The proposition was simple: a people-based process seemed like the single best approach he had not yet tried for solving his organization's growing challenge. It offered an answer not only for pulling people together, but for reducing variance from restaurant to restaurant.

To Mitzel the claims looked like a big stretch so he set some demanding targets: he wanted to enjoy additional productivity advantages that would slash costs and improve profitability, while at the same time not sacrifice food quality or service. Additionally, he knew he would count the effort a failure if in the chase after business he lost sight of a value he had held since starting the company: treating employees as the most valued asset.

John knew that the hard business measures presented no problem—dollar volume per person and per table, as well as shrinkage, retention of staff, overtime, absenteeism, and the rest. But how could they measure the wellness of a workplace? How do you measure employees' perceptions of their work environment and relate this seemingly intangible valuation to bottom-line improvements? And would it be possible to show a correlation *in each restaurant* between an improved workplace environment and improved profitability? Would it really make a difference to the bottom line of an operation . . . or was this Shared Values approach just "feel-good" stuff—the flip side of the kick-'em-in-the-butt, Attila the Hun assaults popularized by some of the business gurus?

Mitzel wanted the effort to hit the bull's-eye on a dual target—pull people together, and also build profits so that his company could improve its existing restaurants and forge a strategy for expansion. He knew that if Shared Values succeeded, he would gain a strategic advantage over every competitor.

When an effort like installing Shared Values lacks unremitting pressure from corporate leadership and senior managers, it's likely to founder. By tracking the impact of Shared Values on each of the eight restaurants separately, the change process would also provide an opportunity to see how the difference in commitment of the individual restaurant managers would influence the outcome.

This was an exciting and heady possibility. But as a realist and experienced businessman, Mitzel knew it wouldn't be easy.

The late comedian Groucho Marx had a classic line about just such a defining moment. In one of the Marx Brothers' movies, Groucho was asked whether he was a man or a mouse, and he answered, "Throw a piece of cheese on the floor and let's find out!"

The cheese thrown on the floor by John Mitzel was the challenge for us to prove that the Shared Values approach could work. He set some difficult criteria for our effort: after the Shared Values approach had had time to take hold, each restaurant would be ranked from Best to Worst in how the employees rated their work environment; and each would be ranked against a financial measure that Mitzel himself would devise. Then he'd compare the lists; if we had succeeded, the level of Shared Values would correlate to Mitzel's home-designed financial barometer.

It was from this initial challenge and the courage of a very unique entrepreneur to experiment that the Values & Attitude Study was created.

THE VALUES & ATTITUDE STUDY

Until that time, many leading psychologists, organizational development experts, academics, and business leaders had been convinced that treating people well was the correct business decision and would lead to increased success in the marketplace—but no organization or individual had ever successfully correlated employees' attitudes with the hard empirical data of how human feelings and attitude impacted the bottom line.

When John Mitzel volunteered his organization as a research lab, it gave us the opportunity of a lifetime. The challenge was significant: we would have to construct a scientific tool to measure how managers, employees, and individual contributors felt they were being treated. Only then could a clear correlation between this measure of employee response and the organization's financial performance be analyzed.

The restaurant chain turned out to be the perfect test case from several perspectives. Each restaurant functioned as a

stand-alone operation with an individual monthly Profit and
Loss statement. Each offered the identical menu and prices,
and each was situated along the I-5 corridor of western Washington, spaced about 20 to 50 miles apart—so the customer de-
mographics were very similar at all locations. This was a test
made in heaven, as some researchers have observed in hindsight. In fact, the only difference was the people: eight sites,
each with different managers and different employees. And
that was what made it the perfect test. The study was divided
into three groupings (see Table 1.1).

Table 1.1 VALUES & ATTITUDE STUDY

GROUP ONE Heroic Principles 8 Shared Values	GROUP TWO Job Satisfaction 9 Satisfaction Levels	GROUP THREE People Systems and Processes 10 Systemic Issues
Monitoring Shared Values	Monitoring individual satisfaction levels	Monitoring opinions about the organization's people-systems standards, infrastructure, and processes
• Honesty—individual and organizational • Truthfulness—individual and management • Trust • Openness to new ideas regardless of the origin • Encouragement to take risk • Giving credit for new ideas and a good job • Putting the interests of others first • Mentoring—individual and management	• Having control over one's work • Believing fairness exists • Having fun on the job • Feeling valued by co-workers • Feeling accepted by co-workers • Feeling well informed on important issues • Feeling trusted by management • Enjoying consistent and believable management • Having Pride in the Organization	• Product quality • Service levels • Ethics • Leadership • Hiring practices • Appraisal and evaluation • Compensation • Promotional opportunity • Communication and self-expression • New employee orientation

Group One asked questions about the eight basic Shared Values that we had isolated from the 17 million surveyed participants in the University of Chicago study.

Group Two addressed Job Satisfiers; the chart identifies the nine satisfaction measures that we've been collecting since 1989. (Today, we have accumulated data to establish international benchmarks in each of the nine areas of Group Two, and we have also compiled what we call *World-Class Scores*— based on results achieved by top-performing companies, that are used as a standard for judging other companies.)

Groups One and Two have turned out to be *the best predictors of future profitability*—a statement that may cause some raised eyebrows, but that we have the data to prove.

Group Three addresses the People Systems and existing processes within an organization (hiring practices, compensation, and so forth). These measurements can lead the organization to find ways of improving the bureaucratic and process elements of its work environment, and increasing the linkages between an organization and its people. Tracking these measures can make every new strategic initiative easier to execute and sustain.

(The Group Two measures also turn out to be valuable indicators to a company considering an expansion or merger. And the numbers from Group One and Group Two are highly valuable to have before investing in a company because of their numeric correlation to future success—true not just for the individual stock market investor but for the venture capitalist and investment banker, as well; if past performance were a valid indicator of future success, every student of the stock and bond market would be rich!)

THE VALUE TENSION INDEX

The ingenious part of the design for the Values & Attitude Study, however, is based on an element of comparison. For each of the value items, every manager or employee is instructed to assign a rating from 1 to 10 on two separate perspectives:

- How important they perceive the value is to them
- How successful the organization is in delivering this Shared Value to them

For example: How important is *trust* to you, and how well is the organization delivering that trust to you and your colleagues? The significance lies not so much in the raw numbers, but in how much the numbers differ. If the employees (again, this is managers as well as staff) on average report trust as being very important to them, and they see the organization delivering a sufficient amount of trust, then the gap would be narrow and the numerical score would be small. We call this gap the *Value Tension Index,* or VTI Score.

RESULTS OF THE I-5 CORRIDOR PROJECT

Benchmarking gives a study like this one its true value. For the restaurant chain, benchmarking took the form of a before-and-after: one study done before the formal Shared Values approach was introduced, and another nearly two years later—long enough to measure not only the impact of a focus on a Shared Values philosophy and the improvement in the ways people are treated, but also the residual effects after the dust had settled.

Overall, the company showed a 32 percent improvement in the Value Tension Index in the two years between the initial Values & Attitude Study and the follow-up (a reduction in VTI from 29.5 to 20.0), demonstrating a significant improvement in the workplace environment and the success of the Shared Values approach.

But what was the impact on profitability? Here the difference among the eight business units was particularly striking.

Comparisons were made between the improvement in employee attitudes about their environment, and business performance improvement at their restaurant. Business performance was judged on three factors, using the index devised by Mitzel himself:

- Net income before operating expenses
- Performance against plan (quality levels and consistency; shrinkage; per-check averages; cost of goods; damages; absenteeism; turnover; and meeting internal audits on cleanliness)
- Profitability (quality levels and consistency; shrinkage; per-check averages; cost of goods; and damages)

The table below shows how the eight restaurants ranked after two years (with individual restaurants designated as A through H).

Performance of the Eight Restaurants Two Years After Beginning the Shared Values Approach

PERFORMANCE	POSITIVE WORK ENVIRONMENT Rating by employees on their work environment (the "People Values")	FINANCIAL PERFORMANCE Net Income Before Operating Expenses, Performance Against Plan, Profitability
Best	A	B
	B	A
	C	C
	D	E
	E	D
	F	F
	G	G
Worst	H	H

The manager of restaurant A and his people achieved the highest ratings for their work environment, and the second best profitability overall.

This was especially surprising to company management,

since A was the newest facility in the chain, open for only seven months when the second testing was done. Where previously no new restaurant had ever become profitable before the second year, restaurant A was in the black in about six months. The superior performance compared to the other restaurants may have been due in part to new employees at restaurant A being hired on the basis of the principles of Shared Values; also, the restaurant was the only one where Shared Values was introduced to all employees from day one of operations.

As in golf, the hope is to achieve a VTI Score (that is, the gap between what people want and what they perceive they are getting) as *low* as possible. In the case of the Mitzel restaurant chain, and in over 1,200 work groups tested since, the lower the VTI Score, the more financially successful the site and the *higher* the VTI Score, the worse the financial performance.

BEYOND THE EIGHT VALUES

Just as the Bill Clinton 1992 campaign staff would later hang a sign on the wall of its headquarters, "It's the Economy, Stupid!" years earlier we had nailed a sign to our wall:

You can't change people, but you can change the context.

Context, the workplace environment—in particular, the way people relate to one another—was, we came to understand, the idea that differentiated the Shared Values approach from everything that had gone before: we were not trying to *change* people by training them, à la the skills-training approach growing out of the work of Frederick Taylor, nor were we were trying to *motivate, improve,* or *change* people in the manner of B. F. Skinner, Stephen Covey, and Tony Robbins (for a discussion of the three generations of employee training, see Chapter 3).

Instead of changing people, we were changing the environment they worked in, making it one that allowed people to function more effectively, with greater confidence and success. Shared Values was providing the mortar that cements universal human needs and aspirations to organizational goals and business strategies.

Yet what of the values about how people deal with other people? Over time, we came to see the need for several additional models of social interaction:

- Values-based consensus building
- Personal responsibility-taking and the inalienable rights of the individual
- Values-based decision-making

We came to call the need for these elements in a work environment "psychic income," as opposed to financial income. And this income, as you might expect, can't be faked, incentivized, reengineered, or superimposed on the employees, managers, or organization.

Psychic income—to grasp one's share of the dream of true fulfillment— is the deep-seated need in every employee, from the elite members in their corner offices to workers on the shop floor. No oversized office or unexpected promotion can create the peace of mind that people seek so desperately. This searching manifests itself in an alarming statistic: according to a study reported in the *Wall Street Journal,* over one-third of the people who work in the typical company would willingly move to another state for a job that would afford them more peace of mind. Not for money, not for status, but for peace of mind.

A Gallup Poll found that workers ranked good health insurance and other personally interpreted benefits such as interesting work and job security as most important. The same poll also asked whether the person was satisfied by their current job on 11 different items, including the opportunity to learn new skills, being able to work independently, and receiving recognition from co-workers. Fewer than half the people responding said they were receiving job satisfaction from *any* item on the list.

Robert Levering, in his book, *The 100 Best Companies to Work for in America,* also found that people were searching for something more than a paycheck. And from the organization's perspective, Levering concluded that the 100 best U.S. companies first look to see whether a job applicant fits the organization's values, and only afterward focus on a specific slot to fill.

THE VALUES & ATTITUDE QUIZ

The Values & Attitude Quiz on pages 15–16 is a short form of the evaluation study we first developed for the Mitzel restaurant chain, and have since expanded and modified, to evaluate people's attitudes toward their work environment and their satisfaction with that environment. It will be revealing if you take a few minutes to work through the quiz at this point. Even more beneficial: ask a dozen or a few dozen of your co-workers to take the quiz as well, then average the results to provide a view of how a wider group of people judge the work environment of your organization. You may be significantly surprised.

Values & Attitude Quiz: Do You Work in a Heroic Environment?
    Answer the following questions about how important each value is to you in enjoying a productive and meaningful relationship in your work environment and with your co-workers.

Personal needs
How important are the following qualities to you personally?
Not Important   Somewhat Important   Important   Very Important

Honesty: Having honesty among the organization's employees.

  0   1   2   3   4   5   6   7   8   9   10

Truth: Always telling the truth within the organization.

  0   1   2   3   4   5   6   7   8   9   10

Trust: Trusting the employees.

  0   1   2   3   4   5   6   7   8   9   10

New ideas: Being open and receptive to new ideas.

  0   1   2   3   4   5   6   7   8   9   10

Risk-taking: Taking the risk to present your ideas or beliefs even if not everyone agrees.

  0   1   2   3   4   5   6   7   8   9   10

Giving credit: Giving credit where credit is due.

  0   1   2   3   4   5   6   7   8   9   10

Selfless behavior: Putting the interests of others first.

  0   1   2   3   4   5   6   7   8   9   10

Mentoring: Taking the time needed to teach and help others.

  0   1   2   3   4   5   6   7   8   9   10

Add up your score (count circled numbers): _____.

Next, divide your score by 8. Your Personal Needs number is: _____ (A).

Your perceptions of your work environment
Evaluate your organization as you see it *today.*
Very Poor   Poor   Good   Excellent

Honesty: Having honesty among the organization's employees.

  0   1   2   3   4   5   6   7   8   9   10

Truth: Always telling the truth within the organization.

  0   1   2   3   4   5   6   7   8   9   10

Trust: Trusting the employees.

  0   1   2   3   4   5   6   7   8   9   10

New Ideas: Being open and receptive to new ideas.

  0   1   2   3   4   5   6   7   8   9   10

Risk-taking: Taking the risk to present your ideas or beliefs even if not everyone agrees.

  0   1   2   3   4   5   6   7   8   9   10

Giving credit: Giving credit where credit is due.

  0   1   2   3   4   5   6   7   8   9   10

Selfless behavior: Putting the interests of others first.

  0   1   2   3   4   5   6   7   8   9   10

Mentoring: Taking the time needed to teach and help others.

  0   1   2   3   4   5   6   7   8   9   10

Add up your score (count circled numbers): _____.

Next, divide your score by 8. Your Perception of Your Work Environment number is: _____(B).

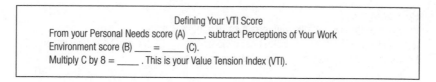

> Defining Your VTI Score
>
> From your Personal Needs score (A) ___, subtract Perceptions of Your Work Environment score (B) ___ = ___ (C).
>
> Multiply C by 8 = ____ . This is your Value Tension Index (VTI).

Value Tension Index Analysis

Building a Heroic Environment requires that everyone shares the same values. The following benchmarks are based on the 17 million surveys, and appear to be accurate on all four continents. For assistance in interpreting results, call (800) 423-9327 in the United States, or fax (425) 828-3552. Or access the Web site at: http://sharedvalues.com

The following assumes that you have obtained average VTI Scores from a number of people in your organization; if the scores you are using are only from your own quiz results, the write-ups reflect your own personal evaluation of your work environment.

Above-Average VTI Scores

6-11	Scores in this range suggest a fun place to work where people are respected for their diversity. You will normally have a harmonious group of co-workers who share your values. High levels of productivity are enjoyed. People are listened to, leadership is shared, and creativity flourishes.
12-16	Your company is a nice place to work. Several of your values are not completely shared, but frank and open conversations about the eight values will provide some healthy perspective on each value. Making requests and not registering complaints will be a productive approach to reducing the VTI score in the future. Leaps in productivity and team spirit are possible.

Mid-Range VTI Scores

17-21	This is the "swing" category. Organizations that launch a Shared Values Process can witness improvements of 30 percent or more in the reduction of the VTI. A good place to begin: build consensus around a willingness to experiment for a certain period of time.
22-26	Tom Peters has suggested that most businesses run their operation at a C-grade level. He was talking about the work environment as well as the productivity level. Scores in this range suggest that people who are doers and motivators, people willing to take the initiative, have been silenced.

Dangerously High VTI Scores

27-31	The organization is in jeopardy of imploding. Turnover may be higher than you like. Retaining good people is difficult and the customer is not being well served. Astonishing opportunities await you if you can turn the VTI numbers around.
32-37	If you scored your organization in this range, you are probably already looking for another job. The organization has enough Dissidents, Fallen Heroes, and Mavericks to start a revolution.

SO MUCH FOR PEOPLE VALUES; WHAT ABOUT BUSINESS VALUES?

We came early to an understanding of the role of People Values. But what about the linkage that connects people and their values with the values that define the organization itself and the way the organization behaves with respect to the outside world—its marketplace? These other values that define how the organization operates we call the Business Values. (We capitalize this term when referring to our specific approach, to distinguish from "business values" as a general concept.)

Business Values define how the organization and its people function. They are the principles upon which everyone in the organization operates, from the chimney sweep to the chairman. Business Values clarify who we are, what we stand for, and why we do business the way we do. They are the basic values successful organizations adhere to, while People Values are those basic human issues that affect daily work performance such as honesty, truthfulness, trust, risk-taking, and receptivity to new ideas. In a well-functioning organization, these two are always in balance. (See the illustration on page 18.)

Like People Values, every company already has a set of Business Values, even though the employees, and even the management, are usually not aware of them; ask any dozen people in a typical organization and you will get a dozen different answers. Ask a dozen senior executives and you will get just as much variation. Yet Business Values resonate within the "context" of every operation.

There are five key elements that a Business Value must satisfy:

- A Business Value must *affect* everything within an organization. It cannot be related to just one department, group, or regional area.

ORGANIZATION-WIDE VALUES

PEOPLE
VALUES

BUSINESS
VALUES

Business Values and People Values in balance.

- A Business Value must be *linked* to the overall success of the organization.
- A Business Value must be *controllable* by something or someone, internal or external. Reputation, for example, is a view held by others, and can be influenced but not controlled. "Selling only quality products" can be a Business Value, because the company can control its product quality.
- A Business Value must be *quantifiable.* This may be done by measuring customer satisfaction, on-time delivery, product returns, or the like.
- A Business Value must be *aspirational* to those who will be involved with the organization. A wonderful example is Nike's "Just Do It," which implies many aspirational elements. To an outsider, the phrase means something much different than to the shipping staff member who needs to finish by today's shipping deadline—yet both benefit from the clarity of the statement. "Just Do It" carries a much stronger meaning than just those three words. (Caution needs to be taken, though, that a great aspirational phrase serves as more than just an advertising slogan.)

THE OPERATING SYSTEM

Sometimes it takes the objective view of an outsider to help people grasp the impact of their own ideas. While still formulating the Shared Values ideas, we discussed our research with Professor Joe Lipson, an exceptional scholar from the University of California, then on loan to the Educational Testing Service in Princeton, New Jersey. After listening attentively, he replied in a kind but triumphant tone, "Do you know what you are really doing? You've discovered how to give people in organizations the *permission* to act in a way they always wanted to act. In essence, you are allowing people the choice to become *heroic.*"

Heroic not in terms of the physical bravery of a firefighter or soldier, but in terms of giving people in the workplace the gift of liberation from their frustrations. We were giving people their organizational freedom.

The simple phrase, "giving people permission to act heroically" has echoed through all our efforts ever since.

Lipson's advice fit well with something our group had recognized early on: you cannot *tell* people to adopt these eight values, as though they needed a lecture or would even respond to one. People already had some understanding of what it could mean to share values—an intuitive understanding inside them. But institutionalizing values to support everything an organization does would add power to the work environment without losing the human qualities.

Two things well worth repeating for both your organization and your family: don't try to *fix* or *change* people . . . and focus on the *context* of the work environment.

What we came to realize was that the most effective way to grasp the context of a work environment was to look at the organization's "Operating System." This term had previously been used only by computer techno-weenies, referring to the instructions that tell a computer how to operate—how to organize data files, display a memo on the screen, divide 376 by 281.

In the same way, a company's operating system defines how the company operates, formally and informally, in terms of organization, structure, and behavior. It encompasses everything from how the employees deal with a customer complaint, to how they respond to an infraction by a fellow employee.

Every organization has an operating system—the company's belief about its people, and about the role of the organization in the larger world. It is an organization's understanding of the reality of its world, encompassing its fears, hopes, vision, mission, business objectives, goals, and frustrations. The operating system is the observable reality of the organization's view of the world—including people, customers, business, and government relations. Any organization not already clear about its operating system needs to examine it, define it, and express it in the simplest terms for all employees.

One business leader turned business-book writer, Ralph Stayer, defined an operating system for his company (though he didn't call it that); it ran like this:

1. **People want to be great.**
   If they aren't, it's because management won't let them.
2. **Performance begins with each individual's expectations.**
   Influence what people expect and you influence how people perform.
3. **Expectations are driven partly by goals, vision, symbols, and semantics, and partly by the context in which people work.**
   That is, by such things as compensation systems, production practices, and decision-making structures.
4. **The actions of managers shape expectations.**
5. **Learning is a process, not a goal.**
   Each new insight creates a new layer of potential insight.
6. **The organization's results reflect the CEO and his or her performance.**

If I want to change the results, I have to change myself first. This is particularly true for the . . . CEO, but it is equally true for every employee. [By "change," Stayer was referring to a change in the way of thinking about people, not a change in style.]

CONTEXT

But how does an Operating System work? An Operating System is the link between an operation and the people. Another way of stating this is with a term we have already used: "context." The context comprises the surrounding elements, circumstances, icons, standards, paradigms and control systems, the policies and procedures that allow people to do certain things and impede them from doing other things. In short, it is the belief system.

It's the context of a situation which dictates that we may scream at a horror movie, a ball game, a rock concert, perhaps at a parade, but not at a business meeting or a dinner party. We may laugh during a wedding reception, but not during the wedding itself. In the right circumstances we may cry during a funeral, horror movie, rock concert, wedding . . . but not at work. Even women, who have a much greater latitude in this society to cry than do men, can rarely cry at work with impunity.

In the same way that a person's behavior is linked through the specific "context" imposed upon him or her at any one time, so the behavior of a person within an organization is imposed by the context of that organization.

Context can also be taken to mean the culture of the company, plus how things are done and how things are not done, what things are done and what are not. It deals, not with what is true, but with how we tell the truth. Context is the invisible link between the individual and the organization. At some companies, the culture prizes truth. In others, where managers only want to hear good news, they will build a context that discourages the sharing of accurate information. "Shooting the messenger" is an anti-truth context.

As individuals we have contexts based on our personal and familial belief systems, contexts about who we are, what we do, how we work, play, and make friends. In the workplace, an individual will behave in a specific way depending on the context of the environment. The new employee very quickly learns to mirror the beliefs and behaviors of the organization.

Paradigms are invisible; context is decidedly visible from the behavior of people. The people of an organization can, in a conscious way, examine their context—and understand how it has influenced their behavior.

Just as there are laws of physics, so are there laws of dealing with people; and people who are in touch with their context can make easier and more appropriate decisions when difficult problems arise.

THE POWER OF CONTEXT

The beauty of using the concept of context to leverage improvements is that once a new context is defined and embraced (and that's the challenging part), an organization takes hold of its own dynamics in a sustainable way—precisely as Professor Lipson had envisioned.

The context of an organization has to be centered around customers' needs and support processes—not around the convenience of any hierarchy, department, division, plant, or group. In a Shared Values environment, there's no need to "flatten" the organization by getting rid of managers; instead, we redefine their function: managers take on the new, more productive role of coach/mentor/resource and, a role beyond that, as "Wise Counsel" (Chapter 12). In Shared Values, there are only two types of players: front-line people, and people who serve as resources to the front-line people. After all, if the current version of management was on target, why would productivity so often go up when the boss goes on a two-week vacation?

As we'll see later, the employees themselves can learn to lead the organization. Radical notion! *Exciting* notion!

EXAMPLES OF ORGANIZATIONAL CONTEXT

When co-author Rob Lebow was 15 years old, he went to buy his first 35mm camera, heading for the store that was the first major discounter of cameras and consumer electronics—47th Street Photo in Manhattan, at the time the only operation of its kind in the nation. After young Rob had worked his way through the clogged aisles and then waited patiently in line at the camera counter for an hour, his turn finally came.

The clerk looked down at him and shouted, "Yeah?"

"I have some questions," Rob responded.

"Next!" shouted the clerk.

It was Rob's first introduction to what he would later recognize as the concept of context: when you wanted to buy at 47th Street Photo, you got your questions answered from a magazine, from friends, from somewhere, anywhere, but you didn't bring them through the door at the discounter's. The store's context was based on volume; a clerk who stopped to answer questions would have been flying in the face of the context, and could expect to bring the wrath of the store manager down on his head. Business Values are sometimes hard-edged. (Today, the world of the discounter has been turned on its ear by chains such as Eagle Hardware, Home Depot, and Office Depot, which actually add service and other types of consumer appeal to the attraction of lower prices: discounting with a new context.)

Campers, hikers, hunters, and backpackers have long known that if they wanted a lantern they could rely on, they had only to buy a Coleman.

The story is told about a day when old man Coleman, the founder of the company, showed up, even though he was retired, and dropped in on a production meeting. He arrived in time to hear a discussion about a defective gas jet in one of the products that leaked under pressure. The question: should the company, famed for its impeccable quality, publicly admit the defect and launch a product recall?

The problem: several hundred thousand units were already on store shelves for the camping season, and only a small proportion were thought to have the problem. Mr. Coleman listened quietly to the discussion about the pros and cons of a recall. "There are so few bad ones," a marketing person reasoned, "we could just wait till people send them back— after all, think of the cost of a total recall!"

That was too much for Mr. Coleman. Red-faced, he slammed the table and demanded, "What's the matter with you people. Don't you know that Coleman lanterns *have* to work?!"

To the company founder, the values were clear: a defective product left no room for discussion or negotiation. The context of the company he had built demanded products that would be thoroughly dependable and stand up in use. The relationship to the customer had to be held sacred.

In the early 1980s, Cadillac stopped being Cadillac, a benchmark of automotive excellence. It happened when the company began using engines from other GM cars and when it sold cars with diesel engines that were something less than buyers had come to expect. What had once been the car people aspired to when they achieved success became instead the car for mature drivers who needed interior space, seating comfort, and the illusion of luxury. BMW, Lexus, and Mercedes were able to take over Cadillac's market share and were able to maintain their promises and their contexts; the comeback for Cadillac was an uphill struggle.

Once a company compromises its promise, its context alters. When a company's employees observe that the organization no longer maintains high values, they rapidly lose pride, and the company loses the edge in the market. We regularly see examples in Detroit, Silicon Valley, and Washington, D.C.

For better or worse, the context of a company can be changed, even by an external advertising campaign. When United Airlines launched its "Fly the Friendly Skies . . ." campaign, the levels of service went up, and revenues followed skyward. The employees hadn't changed; they simply adapted, in a positive, "upward" way. But it works in the other direction as well: after the campaign ended, United fell back into the pack.

Children know who among them is vandalizing their school, but don't feel responsible to do anything about it. By the same token, when the states took responsibility for the schools and told parents to keep hands off, quality of education started going down and the schools have never been the same. It remains true that many top schools are the ones most actively supported by involved parents.

A shining example of responsibility at work in the classroom is the Olympus Northwest School, a small (110 students) middle school in Bellevue, Washington. All important decisions on curriculum and running the operation are made by a group comprised of three staff members, six parents, and six students. Head teacher/principal Jan Fluter points out that it's virtually unheard of for staff to be outnumbered, but appropriate at Olympus where taking responsibility is seen as a defining issue. Fluter says of her 12- to 14-year-old charges, "They can act like age two today, and age 20 tomorrow."

Typical of the responsibility placed on the youngsters: the school is currently struggling with the question of whether to install lockers; there's no budget for it, so the funds would have to come out of another program or the kids would have to organize and run a fund-raiser. But it's the students themselves who will make the decision. On issues like this, each youngster gets to vote one of three ways: thumbs up, "I support the idea"; palm down, "I'm willing to go along with it"; or thumbs down. But they can't simply be against: anyone voting thumbs down is expected to present an alternative suggestion. Still, the power of thumbs down is compelling: no action moves forward until not a single student votes against. "It takes longer to reach a decision," Ms. Fluter admits. "But once we do, nobody goes around griping, 'That's a stupid idea,' because they had their chance to speak up and influence the outcome."

Another familiar place where the power of context can be observed: the difference between how union and nonunion workers behave within one company. The simple phrase "an honest day's work" means something different to the two groups. To the typical union employee it means showing up on time, taking work breaks and lunch breaks under contracted provisions, federally endorsed and legally binding, with anything over eight hours paid at time-and-a-half. You can almost imagine his response if challenged, the same response that Robert De Niro had in the film *Taxi Driver:* "You talkin' to *me!?*"

While in some special situations this context is, unfortunately, as true as it was 70 years ago, in many organizations the work attitudes persist even though management attitudes have greatly improved. But management in most companies has done little or nothing to change the perception of the context.

Is your organization in control of its context, or is tradition governing everyone's actions? "This is the police department, this is how we think"; "This is a railroad, we've always done it this way." Don't blame your employees if you have done nothing to change the context of your operation or if you are doing nothing to inform your people about the changes you've made.

MELDING SOCIAL PSYCHOLOGY WITH ORGANIZATIONAL DEVELOPMENT

If context is a powerful tool for transformation, then how do you go about taking control of your organization's context? Students of business theory are familiar with two traditional approaches to influencing organizational behavior: social psychology and organizational development.

Social psychology approaches the problem by looking at how people relate, behave, and act in groups. The organizational development approach treats the organization as an organism or institution, and analyzes the issues of structure and process.

Few, if any, organizations have harnessed the true potential of organizational development to improve performance and productivity. It has been equally difficult to meld the social psychology elements of a work environment to the overriding processes and change agents. And the most difficult aspect of all, consistently melding individual goals and aspirations to organizational goals, has eluded even the best-managed enterprises. Yet this melding is the secret power to the Process' success.

The Shared Values Process approach doesn't subscribe to either of these competing schools of thought as the dominant solution. Rather, we meld both approaches and harness them jointly to the challenge of taking control of an organization's context. The chart on page 28 represents the interrelationship of these elements in the Shared Values environment.

SHARED VALUES IN ACTION

So much for the theory. What changes has Shared Values been bringing to the workplace environment?

- Long-standing, unspoken hypocrisies that have never been challenged because of fear of retribution are finally being faced. A housecleaning of standards of behavior, not people, takes place. People at all levels of the organization look each other straight in the eye and use the guidelines of *uncompromising truth* to sort through the newly recognized maze of dysfunctional issues and procedures.

Shared Values Process® Operating System

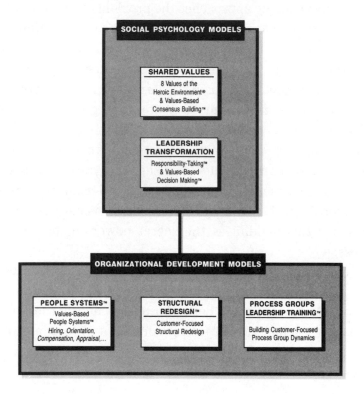

- For the first time people purposefully and carefully challenge the context and recraft their work environment, instead of placing blame. It isn't that the problems and frustrations all go away; the significant change is how people at all levels take charge of addressing the problems using the new approach. Constructive conversations replace "politically correct" empire building. Denial behaviors—refusing to see things as they really are— become replaced with behaviors of recognition, respect, and proactive taking of responsibility. Instead of assigning blame, people dig to find root causes, and search for long-term strategies to eliminate the problem. New standards are set that create resonating values for everyone.

* Managers speak of a revolution centered around values and respect for people.

SHARED VALUES IMPACT ON THE BOTTOM LINE

From a management viewpoint, the ultimate issue about Shared Values lies in asking whether it will have a demonstrable impact on the company's bottom line.

The Pepsi-Cola story at the beginning of this book gives one clear answer: the Pepsi bottler could have faced the kind of disaster that has sent other companies into bankruptcy; instead, guided by Shared Values, the company came through the difficult incident earning praise for its actions and with no damage to its reputation or its profits.

The experience of the restaurant chain cited in this chapter showed the direct correlation between the Shared Values process and improved profitability. In almost every instance, the relative success of a restaurant in embracing the Shared Values approach exactly matched that restaurant's relative success in profitability improvements.

And that's the experience of the companies we have worked with over the last ten years. Every one of these organizations has experienced greater financial success—some merely improving, others improving explosively, within 18 to 24 months of introducing the Shared Values Process. (Probably the most extreme example was a firm listed on the NASDAQ exchange which saw its stock price soar from $1.16 to over $16 per share within three years of launching its Shared Values effort.)

WRAP UP

Without a process that brings a new values-based context to the workplace, little hope exists of retaining the organization's very best employees, managers, and leaders—for they, like us, will be on a never-ending search for peace of mind in the next "greener pasture."

Does Shared Values require deep commitment and significant effort to install within an organization? You bet it does. But the transformation is in itself a true pleasure for everyone involved. Marshall McLuhan said the medium is the message; we say the transformation is the pleasure.

It's no slam dunk. It takes commitment, hard work, personal patience, miles of forgiveness, and tons of personal courage.

People want to be great. The Shared Values philosophy creates an environment where that can happen.

Is it worth the effort? Without doubt. Because the effort leads to an environment where people, at last, can work together effectively, supportively, rewardingly . . . an environment that produces *lasting change.*

2

What We Allow, We Teach

When I see a thick employee manual, I know I'm looking at a
slow company, one that's struggling under a lot of "Halt, who goes
there"-ism and excess baggage.

Tom Peters, in *The Pursuit of Wow!*

SILENT LESSONS

Johnny Waterhands (a fictitious name, of course) had been
diligently scrubbing glassware in the research lab of a
major pharmaceutical company for more than 15 years. A
soft-spoken man with short-cropped hair and an easy, slightly
crooked smile, Johnny was a familiar figure to the scientists,
who acknowledged with the briefest of automatic nods his
twice-daily visits to pick up the dirties and return the cleans,
but who otherwise took him for granted.

So did management. This was a company with a renowned,
much-applauded pay policy that rewarded all employees, even
the hourly workers, with regular increases based on longevity.
Johnny's pay after so many years, while meager when compared even to a newly hired lab worker just out of school,
seemed like a lot of money for so humble a job. In one of those

too-familiar budget crunches, his job fell to the ax. Johnny was replaced by a younger man at a lower salary.

The new employee worked just as hard, seemed just as industrious, ran the water just as hot, left items in the acid bath just as long, yet time after time the scientists would find their experiments spoiled. No head-scratching over this one: the problems had started when Johnny left. Despite repeated training, the new worker just wasn't able to get the glassware truly, thoroughly clean.

You're probably expecting a happy ending—the department head recognized the folly of the decision and brought Johnny back. Only in TV shows and feel-good movies. The lab suffered through several different employees and finally settled on a compromise, a worker not as good as Johnny but at least better than any of the others. The head of the lab called the experience "tougher than hiring a scientist," and things settled back into a routine.

We teach by what we do—the examples we set; that much is obvious.

What isn't so obvious is how many bad lessons we teach by what we *allow*. Every one of our actions sends a message to the people around us about what our values are, what our standards are, what we'll tolerate, and what we consider unacceptable. We do this as parents when we tell our children to be honest and then, when a clerk gives them too much change, fail to send them back to correct the error. We do it as employees when we order a computer game on a company purchase order, have the printer charge a little extra for the brochures in exchange for having him "throw in" a few boxes of personal stationery, or pay somebody from the cleaning crew to wash our car on company time.

When the pharmaceutical company fired Johnny to save money, it sent two messages to everyone in the lab. In effect it announced that doing fine work and being reliable is no guarantee you'll keep your job. And it demonstrated what its choice would be between saving money or saving people—a clear statement of an unfortunate organizational context.

Should the scientists have said something about the dismissal of Johnny? That may be a question for the ages. Perhaps they were afraid to speak up, or perhaps they were just so accustomed to staying out of the fray that they would allow such things to happen without protest.

We send messages every day through obvious and even subtle actions, and in things we allow to happen. The messages we send tell other people what we believe is okay. What we allow, we teach.

Are there zero-tolerance areas within your organization? What happens when an order is going to be shipped later than promised—is there a process for notifying the customer and sales rep? Is there a process for exploring the reasons and providing feedback so the same problem doesn't keep recurring? If you allow undesirable behaviors to be accepted even some of the time, you send a message that it's okay to behave that way all of the time.

At Federal Express, each package is seen as "golden." Four rules govern the handling of every shipment; the rules are worded more or less as follows:

Deliver the golden package on time
Deliver the golden package on time, every time
Deliver the golden package on time, every time, 100% of
    the time
Deliver the golden package on time, every time, 100% of
    the time, always!

Because of this focus, Federal Express has created an image of exceptional consistency and has retained that image for the past 25 years. FedEx uses maximum force to protect, defend, and deliver exceptional service on each golden package. Again, the power of context at work. (Curiously, the context in the Federal Express public relations department appears to have

been influenced by some sinister force working against the interests of the company. The above rules may not be correctly
quoted, but it's virtually impossible to find out the exact language. Requests to the FedEx PR department—for information
about the company, or confirmation on something as public as
what year the current CEO took office, or the correct version of
the four rules—are received with as much enthusiasm as a
bomb threat. Its way of treating every simple request with
grave suspicion and making the caller feel like a criminal and a
snoop demonstrates the challenge in establishing a positive
context throughout all corners of a company.)

What's the action in most companies when a promised delivery date is missed? It ranges from no action at all to shouting at
the production scheduler to leaving a sarcastic voice-mail message for the shipping manager. Or, at best, a drop-everything,
hurry-up effort to get the order finished and on its way . . .
quite probably forcing several other orders to fall behind schedule and so compounding the headache instead of finding an approach that solves the problem in the long term.

During research for this book the authors stayed at the
SoHo Grand Hotel in Manhattan, which has been designed with
an artsy look and aims for a young, with-it crowd. The comment
to a bellman that some of the carpeting was already beginning
to show signs of wear, even though the hotel was only a couple
of years old, brought the reply "That's the way it's supposed to
look. A new guy tried to shine the brass in the lobby, so we [!]
fired him." Apparently, the staff somewhere picked up the idea
that *shabby* was part of the design of the hotel.

In late 1996, a major oil company was accused (and con-
victed by the media) of chest-thumping for their color/
gender/culture-blindness, while executives operated on an unspoken policy that pretty well blocked advancement for anyone
not stamped out of the same mold as themselves.

Americans in all walks of life shook their heads in dismay
over this duplicity. But outside the glare of the cameras, many

companies aren't doing much better. In their cafeterias, day after day, the Latinos all sit together; the Asians have their own tables, unmarked but so well recognized that they might as well carry a sign, "Asians Only." Sure, some Latinos and some Asians eat at the Anglo tables and are accepted on equal terms by their lunch companions. Yet many don't feel comfortable doing that—a sure sign the professed diversity policy isn't a value shared throughout the workforce.

Probably few managers in these companies ever set an example by asking to have lunch at one of the Latino or Asian tables, perhaps in part because it might be perceived as unvarnished tokenism. But by ignoring this informal red-lining in the lunchroom, the company announces that despite the proudly proclaimed policy, it's okay to treat members of minority groups differently. (Take a look around your own company cafeteria. Does it fit this description?)

In the end what counts is the message our actions send. As the end of a quarter nears, sales reps start eyeing their unfilled quotas, nervous because so much of their income depends on reaching or exceeding their goal. Meanwhile the executives fidget over the beating that the stock will take if sales don't meet Wall Street projections. Hyped by this restlessness, it's easy to fall into a practice that's known as "stuffing the channel"—the reps call their good customers and ask a favor: "Let me put in an order for an additional $50,000 of product, and you can ship it back after the beginning of the new quarter." So the quarter ends looking good after all, but the problem compounds itself.

Someone once irreverently described the U.S. president as like a flea on the back of an elephant, because the bureaucracy is so entrenched that a presidential order may simply not get carried out. CEOs sometimes encounter a like situation. The new CEO of one Fortune 500 company found he had an unsaleable several hundred thousand dollars of equipment in the warehouse that had been returned because of stuffing the channel. He took the lumps courageously, announced a write-off of more than half a billion dollars, and vowed it would "never happen again." But the lessons had already been too well learned; the context was too strong. Behind his back, the sales force continued to stuff the channel. Until values change throughout a company, lessons once learned take deep root and fester, and cause the untimely death of too many once-great organizations.

Cited as an example in management books so often it gets almost boring, Nordstrom built its business on shared values. One story old-timers love to tell around the company relates how—in the early days, before the name became famous—the Nordstrom president found himself on an over-full elevator, with people waiting to get on; he stepped off, giving his space to a customer, and walked down three flights of stairs. That's the positive side of teaching by example, and it's the reason that today you'll get the same courteous treatment at Nordstrom whether you're buying a shirt or returning one. The context of its success formula is its aim of building a long-term relationship with each customer.

In Bellevue, Washington, the Acura dealership is owned by a vibrant, dynamic, straight-talking businesswoman, Barbara Nelson Guinn, who once voiced her concern at a staff sales meeting because new car sales were off. "Last month we lost money on new cars," she told the group.

One of the salespeople piped up. "How could that be?" he demanded. "*I* made money." Here again the values were out of sync: the salesman was focused only on how well he personally was doing, overlooking what would happen to his job if the dealership continued to lose money on new-car sales.

This issue of paying salespeople based on how much they sell is so well entrenched that hardly anyone would dare challenge it, yet swirling around the issue is a seething mass of painful emotions. A lot of people in any company, who work long hours and give up other things they would like to be doing out of dedication to their organization, bear a deep-down resentment against sales reps, who work hard, too, yet who know that by putting in an extra 20 hours this week, they'll tuck away a fatter pay envelope.

But how can a company function without commissioned salespeople? If it doesn't seem likely, take the case of Michigan-based Steelcase, the world's largest maker of office furniture, with an enviable market share of 20-plus percent of the sales in their industry. At Steelcase, the field salespeople are on salary like everyone else, not on commission, and that approach has been in force for decades. If the organization succeeds, all employees benefit by way of a profit-sharing plan. The sales reps know they are team members of a larger organization. Steelcase employees will tell you that every member of the team—designers, engineers, purchasing, and manufacturing people, even the shipping and billing clerks—plays a key role in the company's market-share success.

The same attitude takes other forms that run throughout this international market leader. As onetime Steelcase CEO Frank Merlotti described it. "There are no separate departments for different functions anymore. We tried to remove anything that got in the way of people communicating, discussing ideas. We wanted to get rid of this top-down thing."

For another champion of the commission-less sales force, ask Rob Rodin, CEO of Marshall Industries, the California-based electronics distributor. Rodin told journalist Curtis Hartman that a day doesn't go by without someone asking him to defend the pay plan he pressed on the company. "Yes," Rodin said, "I'm the heretic who took the entire sales staff of a $600 million company off commissions, overnight, and lived to tell the tale." Writing in *Fast Company* magazine (June/July 1997), Hartman sums up the result: "[N]one of Marshall's strategic innovations could have happened if the company had not first abolished its decades-old compensation system." Since Rodin became CEO, the article notes, profit per person has tripled, sales per person have more than doubled, and sales revenues have multiplied more than two fold to $1.2 billion—all achieved with *fewer* people on the workforce.

In plain fact, the system of sales quotas and commissions announces to the rest of the workforce that salespeople are making a more important contribution to the success of the firm than the people who dream up the products, build in the quality, or conceive the advertising that makes customers want to buy. But whether any particular company can change a long-established commission system depends on the context of the organization and many other factors.

If the hourly worker says that an honest day's work means putting in a full 9 to 5, what does the professional say? The professional says an honest day's work means "Staying until the job gets done." Both are reasonable answers. Now the hard part: can a workable, healthy corporate culture thrive when these two disparate values live side by side in the same company? What happens is that it becomes too easy to rate anyone who works in the plant as a mere second-class citizen.

When an organization allows differences in the way people are involved—emotionally, physically, financially—it runs the risk of creating a caste system.

Are we going to press you to put all your shop people on an annual salary or to hang time-clocks in the corporate headquarters? No; the solution is quite different, as you will soon discover.

These value issues afflict business, but also abound on a grander scale. Consider the "Buy American" program, with its very noble motives, most laudable intentions. What were we really being asked to do?—buy American-made goods, *regardless of price or quality.* Which is to say, put aside your good judgment, put aside the value system you've always used when shopping, and adopt this other value system instead. Why? Because it's patriotic.

("Buy American" sounded most appealing when it first resounded across the national landscape. But it signaled to our

American manufacturers that they didn't have to play by the same set of rules as the international manufacturers. And after all, the "Buy American" theme wouldn't have been needed if American manufacturers had set values to keep domestic products competitive with imports.)

When, without appropriate repercussions or reprisals, we allow other nations to dump shoes, television sets, or computer chips into our stores below actual market price, we teach them it's okay. We encourage behaviors we soon regret.

In contrast, the Japanese have always taken a protective, jaundiced view of foreign merchandise that could dominate their marketplace; they've placed heavy restrictions on finished goods being imported into their markets. And by also treating basic commodities such as apples, rice, and frozen fish as a threat to their domestic market, they have until recently been successful in protecting their home industries. Helpful to them, but unfair and economically disruptive to the very markets they have taken advantage of.

Under determined outside pressure, the picture is now changing. In hindsight it would have been more effective in our relationship with Japan if we had, from the start, responded with reciprocal trade behaviors and practices. When standards of behavior and fairness are not accurately established, people or nations are encouraged to do as they please, creating a de facto context instead of a Shared Values context.

So if it's clear that employees do a disservice to their organization, regardless of its size, by inappropriate, damaging behaviors, what are the alternatives?

One would be a return to the rigidly hierarchical structures of the past, laying down inflexible rules and demanding that everyone follow them. This lock-step approach to management, a carryover from the old military system where it's not only justified but necessary, no longer meets the needs of a time when product cycles are measured in months instead of years and competitive pressures are so intense that, in Warren Bennis's ghoulish words, "corporations eat each other's flesh and sell off the muscle and bone." The command-and-control structure also worked better in times when people were raised and educated to respond to following orders without question. Today's successful business wants to encourage ideas from its people, not smother their ideas under the blanket of a rigid bureaucracy. But even the military today speaks of its customer; in the non-battlefield situation, it recognizes the change in approach to handling people.

Then if not by laying down the law, how else can an organization teach the right lessons and encourage productive behaviors? The lesson of Mothers Against Drunk Driving (MADD) holds a key to the answer. Instead of a slogan like the ineffectual "Just say no" campaign that probably produced more mirth among teenagers than any desired behavior change, MADD provided a choice example of harnessing the power of context. When it created the brilliant slogan, "Friends don't let friends drive drunk," it was expressing a values-based idea, one that clearly established a context around the positive behavior it wanted to engender. And the slogan goes much deeper than just the preaching of contextual values: it clearly establishes a new action standard for friendship, encouraging friends to be their very best.

How does this statement drill so deep in our heads and hearts? And how does it elicit the expected response? Without prescribing legal parameters for intoxication or suggesting how we should protect our friends from the unthinkable, and without prejudgment, fear, or shame tactics, we are marshaled by this phrase to act heroically. No training, no prescribed policies, no creative brainstorming; instead we are simply called upon to wrestle the keys away from a friend and inconveniently drive out of our way to deliver the person home safely. Few advertising slogans have ever had the emotional power of these six ordinary words.

WRAP UP

By our behavior and the decisions we make, we teach the people around us what we consider acceptable. Every one of our actions sends a message to the people around us about what our values are, what our standards are, what we'll tolerate, and what we consider unacceptable.

But other people pick up messages from us by more than our actions; we also communicate standards by what we allow—all the things that take place around us which we observe and keep silent about.

In a Shared Values environment, employees and managers start to act in ways that teach the people around them what the new standards are, and what behaviors are acceptable.

Your first step toward a new, more fulfilling, more fruitful work environment was taken when you decided to read this book. By recognizing how today's environment leads people in the wrong directions, you can adjust your own behavior in ways that will encourage others to follow, using your behavior as their new standard. Your behavior will create the context.

3

Fixing People
Never Works

If the people don't get it, don't fix the people—fix the process.
W. Edwards Deming

THE PROBLEM

The bright-eyed new employee, walking in from the parking lot, finds himself alongside an older man and says, "This is a pretty big place. About how many people work here?" "Oh, about half," comes the answer.

The joke is familiar and perhaps a trifle shopworn, but it still brings anguished smiles. The real pain, though, isn't over people who don't keep busy and industrious all day; almost all do, especially in this era when the specter of downsizing and layoffs hangs over so many employees. The real issue rears up every time the same problems—overproduction, poor quality, lost files, misdirected shipments—need to be solved all over again.

It rears up every time a blankly smiling co-worker agrees to e-mail the report she completed, but never does; when people sit mutely at a meeting and afterward complain about the decisions; when employees are so focused on their own careers that they can't describe the goals of their group, much less of the overall enterprise.

Since the beginning of the Industrial Age, a pervasive uneasiness has gnawed at executives and managers: all would be rosy if only we could fix the workers. (But you can't fix your mother-in-law, so why do you think it's possible to fix your employees?)

Management practices that used to work don't work anymore. The real problem: many systems and methods of management have outlived their usefulness.

USA Today reflected the current disaffection with employee-management methods in an article (10/17/95) reporting on the 1995 Baldrige National Quality Awards honoring companies that achieve the greatest improvement through their Total Quality Management (TQM) programs. "The quality revolution handed out its medals Monday," the piece began, "but the army is a bit ragtag." Noting that the number of entries had dropped to 47 from a peak of over 100 just four years earlier, writer Martha T. Moore observed that "Companies once gung-ho about quality have become disillusioned . . . and have dropped out."

IT STARTED WITH TAYLOR

The man who gave a kick-start to the modern ideas of management, in the sense of managing a business and managing a workforce, was of course the indomitable Frederick Taylor. His work in the late 1800s–early 1900s set the stage of "scientific" management as he examined the species of employee, manager, and business from virtually every angle—ranging as wide as tool design and incentive pay, work planning and employee selection.

Taylor did his measuring and his timing, and produced his

equations and mathematical models. For him it was all science, all reducible to numbers, treating people as head-count, one of the necessary ingredients of production—as if 20 tons of steel could be made from 40 tons of pig iron, 40 tons of coal, and 5 tons of workers.

But nobody was laughing. Taylor was able to show the skeptics that he could apply his science, examine the work environment, and then increase the efficiency of the workers. By a large margin.

Taylor's insight opened the door for a new occupational group, the industrial engineers, who took as their target the molding of the "successful" employee. From the first they congregated in the factory, focusing their lenses on the production-line worker. The diagnosis: the worker can be *trained*—improved to work better. The initial prescription: for the factory, redesigns for efficient flow; for the worker, large doses of skills training. We now look on the work of Taylor and his disciples as creating the first generation of employee training attitudes.

World War II opened a lot of eyes to new attitudes about the way people work, learn, and respond. Studies done by the military examined issues of stress and fatigue in battle and under pressure. Meanwhile home-front shipyards and factories, in many cases manned (if that's an appropriate word) largely by women, were turning out Liberty ships, tanks, and airplanes at a rate unmatched before and hardly matched since.

All of which gave rise to something called the *human potential movement.* As embraced by the business world, this added a new dimension on top of the skills training. Its theme: let's also try to train the employees in mind-skills. The vocabularies of the mind-skills exponents flowed with phrases like *positive thinking, paradigm shifts, visualization, affirmation, sensitivity training,* and *possibility thinking.*

When the 1960s arrived, this notion of molding human behaviors became the new mantra of top business schools, the

Harvard Business Review, and *Fortune* magazine. Companies continued to train and polish employees' skills and also began to assume responsibility to develop people as team members, managers, and leaders. Two decades later, companies like Motorola, Hewlett-Packard, and Apple Computer were held up as models of this 1960s new-age development movement.

"FIXING" THE EMPLOYEES

While the first-generation ideas sought to train, this second generation seeks to achieve improvement by "fixing" the employees—altering the individual's thinking and frame of mind. Reshape the employee's attitudes and consciousness, the thinking goes, and you can influence the behavior and work performance.

Industrial engineers started to create a science around the parameters of the successful employees in a particular job. Personnel departments changed their name to Human Resources and set their sights on reshaping attitudes and behaviors. And ever since, leading exponents like Zig Ziglar and Stephen Covey have been waving the flag for what might be called "liberating the human potential" through new habits and a new mind-set.

These and other proponents of the human potential movement base their thinking on the work of behaviorists like psychologist B. F. Skinner, emphasizing issues of positive thinking, visualization, affirmation techniques, incentives, and motivations. This behavioralist school of thought became very dominant in business, and persists today in much of the way we "incentivize" and motivate to increase performance.

There's no lack of evidence that the approach can work with startling impact in the short term, but the evidence is also clear that there is little residual, long-term value. A significant amount of research over the past 40 years suggests the reverse of what we've all been taught and what we've come to believe: we don't get better results by dangling incentives in front of people—promotions, salary increases, awards trips. Nor do we intrinsically change people by putting them in front of a video camera or labeling them as "driver-driver" or "amiable."

The bottom line is that manipulation won't change or fix the workers. Perhaps it's not quite fair to say that most corporate development, incentive, quotas, rewards are just forms of manipulation, but it's clear from the continual waves of corporate-change movements—TQM, reengineering, and the rest—that something is still missing.

What is it that's missing?

THE FLAW IN THE SECOND-GENERATION APPROACH

Much of the second-generation effort sprang from analytical and clinical psychology as well as homespun concepts, principles, values, and beliefs. Initial hopes were high because of the motivational aspects of the information and the dynamic way the material was presented. But although compelling and popular, a company investing in the approach eventually discovers that the Achilles' heel of the effort is all too obvious: in large measure, the approach fails to meet the test of sustaining long-term organizational health and growth.

The problem isn't with the ability or the character of the employee, which is why straightforward efforts to solve the problem with information, skills, and "fixing"—trying to alter the individual's thinking and frame of mind—hit a brick wall and fall flat.

A screenwriter once found himself on an airplane sitting next to a television producer, who started trying to talk the writer into doing a pilot for a TV series. The writer had his own projects and wasn't much interested. After a while of getting nowhere, the producer pulled out what he was sure would be the capper: "Don't you realize," he said excitedly, "if you do the pilot and the series goes on the air, you won't even have to write any of the stories—you'll get paid every week and you'll make so much money that you'll never have to write again!"

Writers, artists, and musicians do their thing even when they're not being paid for it. In business, employees come to work in the morning because that's how they earn their money . . . but the money doesn't buy their enthusiasm.

Lakewood Research, the publishers of *Training* magazine, reported on a 1987 study that questioned 3,500 senior managers and HR professionals, seeking to find what percentage of traditional training is retained one year later. The answer respondents gave should have raised a cry of alarm: they reported a number in the range of 10 to 12 percent. And not because the training was poor; the problem lay in poor retention—something that all of us have known or suspected intuitively for years.

In an essay aptly called "Why Change Programs Don't Produce Change," professors Michael Beer, Russell A. Eisenstat, and Bert Spector addressed this problem in the pages of the *Harvard Business Review* (November/December 1990), offering views on why no long-term change occurs from the kind of management approach that depends on individual enlightenment:

> Most change programs don't work because they are guided by a theory of change that is fundamentally flawed. The common belief is that the place to begin is with the knowledge and attitudes of individuals. Changes in attitudes, the theory goes, lead to changes in individual behavior. And changes in individual behavior, repeated by many people, will result in organizational change. According to this model, change is like a conversion experience. Once people "get religion," changes in their behavior will surely follow.
>
> This theory gets the change process exactly backward. In fact, individual behavior is powerfully shaped by the organiza-

tional roles that people play. The most effective way to change behavior, therefore, is to put people into a new organizational context, which imposes new roles, responsibilities, and relationships on them. This creates a situation that, in a sense, "forces" new attitudes and behaviors on people.

This view from the *Harvard Business Review* is supported by our own findings from ten years of work at over 1,200 work sites: it's the establishment of a new context, and a new definition of the manager in a transformational role, that redefines the new organization in a way that brings lasting change.

From this the question becomes, "How can we get people to play at the top of their game, using other than the first-generation skills-based and the second-generation behaviorism-based methods?"

The primary human motivator grows out of what makes most people resonate, which comes neither from developing new skills nor from concepts that the human potential movement deals in. People respond and relate to *values*. And values don't have to be sold to people, because all of us already have a set of fundamental values deeply imbedded; or, as we sometimes phrase it, all people have their values "hardwired" within. Values like trust, loyalty, the need to feel a part of something larger than ourselves are common to virtually all humans, regardless of culture, education, or social class. We have found that there is less diversity within basic values than you would expect.

It's not that the first- and second-generation training approaches were wrong, it's not that skills aren't important. They are, and will continue to be. But with these approaches, the sight is on the wrong goal: "If we could just get people to think the right way . . ."

THE THIRD GENERATION

The answer to "How do we get people to play at the top of their game?" lies in finding a *values*-based approach. This is what we have termed *Third Generation Training*.

What is Third Generation Training? It's the approach we've developed that melds social psychology and organizational development into a new process linking people to their organization in a totally non-manipulative way (see the chart below).

The link or glue in this process is the set of eight Shared Values, values that are readily embraced by everyone regardless of gender, age, experience, national origin, or religion. These eight values, the amalgamation of employees' voices from 40 countries around the world, equip people with what they need to play at the top of their game.

Values are the glue that bonds universal human needs and aspirations to an organization's goals, business strategies, and vision in a process that creates a dynamic balance between Business-Values needs and People-Values needs.

Three Generations of Training

First Generation	Second Generation	Third Generation®
SKILL BASED	HUMAN POTENTIAL	OPERATING SYSTEM
TRAIN PEOPLE IN NEW SKILLS	TRY TO *FIX* PEOPLE'S BEHAVIORS	ESTABLISH A *NEW CONTEXT* IN THE EXISTING ENVIRONMENT
LEARN:	TRAIN IN:	ESTABLISH:
• New Tasks • New Subjects • More & Better Information	• Positive Thinking • Personal Character • Visualization • Affirmations • Aversion Therapy • Sensitivity • Paradigm Shifts • Possibility Thinking	1. SHARED VALUES For The Organization 2. STANDARDS For The Organization 3. SYSTEMS For The Organization 4. STRUCTURES For The Organization

A comparison of first, second, and third generation approaches.

WRAP UP

Companies must still train and update to provide their people with the needed skills, but trying to "fix" the behaviors of people leaves a company stranded. And it leaves the company's people bewildered and discouraged, unprepared to meet the different climates and challenges of the rapidly changing enterprise jungle.

Deming's remark quoted at the beginning of this chapter sums up the point: "If the people don't get it—don't fix the people, fix the process." Even when the system is broken, the people are still capable.

So the Third Generation rule is: stop trying to fix people. Allow people to be great—by fixing the context.

4

People Want to Be Great

When the great management guru W. Edwards Deming met with a group of executives, he would often ask, "How many of you have dead wood on your staff?" Most or all the hands would go up. Deming would then *shout* at the group, "Did you hire them that way, or did you kill them?"

WHERE WE WERE

It used to be that business leaders perceived their employees as lacking imagination and enthusiasm. That disdain applied not just in the shop and on the plant floor, but behind the doors of the offices as well. The requirement for becoming a manager wasn't 20 years of experience, but simply a college education in a particular area of significance to the company. (Bill Gates, who has no college degree, would have been turned away by IBM.) Your educational credentials, not your ability or drive, defined who you were, what you were capable of, and what job you could aspire to.

It's still true that many organizations don't believe most of their people come to work highly motivated to achieve and attain success. Rather, managers still assume that people need to be watched, supervised, and incentivized, that only strong management practices will succeed in herding their people into a mind-set that will produce good results. Too many supervisors and managers still do not believe that the can-do attitude or the creative spirit is inherent within the typical employee. They think that employees will not perform well unless the boss is, as the modern irreverent expression puts it, taking names and kicking butt.

People are relatively neutral, the thinking goes—a resource to be mined—and cannot function effectively, comply with the rules, or meet the prescribed standards without the carrot of materialistic motivation and detailed directions, or the stick of punishment.

The employees, of course, soon enough come to feel that all this is true, that they are essentially incapable of making good decisions on their own and need these structures in order to function successfully. And it's not some vague "them" I'm describing; it's you and me and all of us, except perhaps for the very occasional self-starter, innovator, or entrepreneur—the Steve Jobses, the Donald Trumps, the Sam Waltons.

The notion of people being incapable and needing to be managed is a self-fulfilling prophecy. The American worker came to accept this sense of incompetence as true even though our American heritage contradicts it—the heritage of the men who sailed the whaling ships, the pioneers who settled the West, the immigrants who landed here empty-handed but survived to grow the crops, run the factories, build the cities, and launch the companies.

Our heroes were Horatio Alger and, yes, Andrew Carnegie. So how did the workforce of a nation come to be convinced that it can't succeed unless someone gives it a set of rules and then hangs around to make sure the rules are being followed?

But let's look forward instead of back, and ask not how we got to this sad state, but where we can find examples of what we would like to become, and what we must now do to bring about the kinds of changes that will remake the workplace and become lasting.

STARTING WITH A VISION

In 1961 President John F. Kennedy stirred the American people by announcing the goal of landing a man on the moon and bringing him safely back to Earth before the end of the decade. Intuitively Kennedy sensed that the project demanded a noble goal; he saw that trying to justify the effort with the intent of "beating the Soviets" would have extended the cold war into space. Instead he told the nation, "We do these things not because they are easy but because they are hard"—focusing on the American tradition of hard work.

Americans were fired up by the idea and the press was behind it, but despite this enthusiasm there would be no moon shot unless Congress agreed to provide the funds. NASA, the space agency, put together a business plan, and Senator George Smathers of Florida traveled with a Congressional team to Cape Canaveral (now the Kennedy Space Center) to see the work in progress and hear the plan.

Touring the grounds, Smathers entered one vast, hangar-like room, fully lit but empty except for a cleanup crew. He drew one woman into conversation and asked her what she was doing. "I'm helping get a man to the moon and back safely," she answered.

The senator immediately recognized that NASA not only had a clear vision, but had managed to imbue its employees at every level with it. That was enough to convince him that Kennedy's moon project was possible, and he decided on the spot to give it his support.

In the same way, the reshaping of any organization must start with a clear sense of purpose. When the sign at the end of the tunnel reads "Shared Values," the destination is predefined, and is arrived at by moving all managers and executives away from the long familiar, military-style command-and-control method of operating, to a new model based on the power of people and the amazing magic the collective can achieve.

THE RETURN TO *TRUST*

Some years ago the world's biggest manufacturer of floor maintenance equipment, the Tennant Company, was waging a do-or-die battle against foreign competition. CEO Roger Hale found a way to get rid of the expensive operation geared to fixing defects at the end of the production line: he got employees to start monitoring their own work and setting their own goals to improve quality. The workforce turned into a model of self-management, high morale, and teamwork, capable of creating products that reached the end of the production line not needing any costly fix-it work. No secret to the method: Hale simply put trust in his employees.

Every person in your organization has different talents and different levels of ability: few can design a new product, few can prepare the corporate profit and loss statement, few can write an effective product marketing plan. *But virtually everyone can be performing at a higher level than they currently are.* People individually—and even more so when they act together—have the power to achieve unlimited things. All that's required of us as foremen, supervisors, managers, and executives is allowing people to achieve by giving them our trust.

Trusting people means allowing them to be great.

The whole of the Shared Values philosophy shouts out this one central idea: the belief and conviction that people want to be great.

Why do people leave one company for another? Sure, there are lots of reasons, but a leading one—you may well have had the experience yourself—grows out of the search for a situation where the person will find an opportunity to use his ability, her talents . . . a place where initiative and the willingness to take risks is rewarded with recognition and opportunity. Probably most of the people in your work group or organization are with you after leaving some other company where they were dissatisfied because they were being micromanaged instead of given the opportunity to work at the top of their ability.

How many of today's company leaders left other companies in their quest for psychic income—a sense that they were needed, respected, and trusted? Look what those other companies lost by not providing the right environment and challenges.

> **An organization's greatest responsibility is to create the conditions for superb performance by everyone. It's the job of every manager to allow people to play at the top of their game.**

ESTABLISHING TRUST

If the employees in an organization are not now grabbing for greatness, it's because the beliefs, policies, and procedures of the company and its managers don't let them. Managers need to get out of people's way, and they can only do that by getting rid of outmoded belief structures based on fixing people, on controlling people. When a company embraces a Shared Values approach, it gives people information and then lets them make good choices—a policy we think of as being Jeffersonian in nature.

It's individuals, not organizations, that build nations and make history. And ordinary people have much greater ability to accomplish what needs doing than the current stifling structures allow.

A few true leaders have recognized this principle themselves. The following gives a favorite example.

> The story is told that when Henry Kissinger was Secretary of State, he once asked a staff member to prepare a report on the situation in Chile. When the man brought in the report,

Kissinger, before taking it from him, asked, "Is this the very best you can do?"

The man thought for a moment and then said, "Let me do some more work on it."

When he came back the second time, Kissinger accepted the report. Within a few minutes he called the aide back in and asked the same question: "Is this the very best you can do?" Once again the man took the report back to rework it.

The third time he returned, as he handed over the report he looked Kissinger directly in the eye and said, "This is the best I can do." Kissinger was well satisfied. "Good," he said. "Now I'll read it."

When people are asked for their best, they will let us know when they're achieving it. What they want from leadership is trust. It's remarkable how much power we can bestow on people when we let them know we expect their best.

In the early years of CNN, a news director hired an experienced documentary producer and gave him as a first assignment the difficult job of producing a 30-minute show on air pollution in the local area. When the rough cut was ready and the producer brought it in for review, the news director told him, "I don't want to screen it—I hired you to do good work, there's no need for me to look at it." The message: "I trust you to do quality work so I'm not going to tell you how to do your job." In fact, this attitude permeated all of CNN—the trust that each person would do quality work.

No one has ever encountered an organization where employees threatened to strike because their leaders wouldn't bring in Total Quality Management, or wouldn't downsize. What employees crave is more communication, the sense of feeling respected, the room to make mistakes without getting their heads chopped off. These basic human needs are not satisfied by replacing typewriters with high-powered computers, or by updating procedures manuals, or by loading more authority on the Human Resources people.

One major reason for the success of Microsoft lies in an early decision to keep Human Resources out of most of the hiring. New employee candidates are interviewed by the leader of the group and the peers they'll be working with. Once they've made their decision, HR takes over the paperwork and follow-through. The Microsoft method is another fine example of placing trust in the hands of the right people. And it works.

So what are the secrets to creating an environment where people want to do great things every day? A critical part of the answer lies in giving people access to information, or, in a hot phrase of the day, "opening the books."

OPENING THE BOOKS

If you sat down to design the basic principles for a totalitarian society, you likely would be wily enough to maintain a tight stranglehold on the citizens by depriving them of information. Censor the newspapers, take over the television stations, confiscate the printing presses, shoot the vocal dissenters, and imprison the writers.

Communism collapsed under the weight of its economic failures, but its demise was clearly speeded when the information barriers were toppled by the fax machine, the videocassette, and cable. Once the people living behind the Iron Curtain were able to learn what was really going on, they took their future into their own hands.

Visitors to the Mayan ruins in Mexico find the remnants of what was a remarkably advanced yet gruesome civilization, one based on an accurate knowledge of binary mathematics and astronomy . . . and on human blood sacrifice. With the ability to forecast events such as solar eclipses and the solstices, the leaders constructed buildings so that the priests could, at certain moments in the year, stand in a spot where they would be emblazoned by rays of the sun passing straight through tiny openings in a building hundreds of yards away and lighting them as nobly as gods. But the sharing of math and astronomy would have been seen as blasphemous or worse to the elite, the keepers of the secrets.

The citizens of the modern corporate enterprise have a different use for information: not to overthrow their leaders or undermine them by revealing protected knowledge, but to enhance their individual ability to make a valuable contribution.

Information sets people free—free to do the best work they can. The corporate worker can schedule the length of a meeting by checking on a computer to see how long the last three meetings took; the CAD/CAM operator can check the customer's order to see whether a substitution of one type of alloy steel for another would be acceptable; the loading-dock worker can determine whether a less costly shipping method would get the order to the customer on time.

Information is power. Until the dawn of the Information Age, what separated the manager from the worker was access to the information about the business. It was assumed that the employees had no need or use for the information, and wouldn't know what to do with it if they had it. And besides, there was no practical, efficient way of providing it to them. The personal computer has changed accessibility—creating a situation that is seen as either opportunity or threat. Many managers, unfortunately, view this from the perspective of a Mayan priest. It's as though nothing has changed.

The updated notion of the "open-book manager" seeks to bring the overdue change by letting employees know what's really going on. Because when employees understand where the business is making money and where it's losing money, they will be better employees, able to make better decisions. Open-book management makes stakeholders and partners of employees— something we could have done 100 years ago—a wise decision for the twenty-first century.

But for employees to deserve a stake in the well-being of the organization, they need to understand the financial processes, so the organization needs to be willing to share ongoing details of its financial performance. The employees benefit by greater participation, and the business is rewarded with better decision-making. Employees who interface with customers and are knowledgeable about margins have a greater capacity to negotiate prices, a greater sense of what they can afford to discount but where to draw the line. In manufacturing, if people on the production line understand where money is made and where it's lost, the connections between raw material costs, labor costs, and design and engineering costs, they will conclude that their job extends beyond units per hour and the scrap count. Engineers given the financial data become more sensitive to the expense of manufacturing and much less likely to design items that are too expensive to produce, maintain, and retrofit.

When the books are open, employees discover for themselves where the problems are, where money is leaking out or being wasted. And the peer pressure for improvement becomes enormous.

The idea of opening the books isn't new or original, of course; we offer it as an illustration of the principle that trusting people allows them to be great. An organization willing to share information, especially financial information, and willing to teach how to interpret and apply it, is an organization demonstrating that it will allow its people to be great.

WRAP UP

Modern corporate disciples of the Mayan priests, and those disciples' compliant followers, need to change their philosophy about information, power, partnership, and the potential of people. If you were to put up one sign in your office to remind you of the principles to guide you in your daily decisions, the sign should read:

People Want to Be Great.

The Shared Values effort aims to create an environment where that can happen—here we put up no roadblocks but instead *allow* people to become great.

We have looked in this chapter at the *what* and the *why* of Shared Values; the *how* of making it happen is the topic of the rest of this book.

GETTING PEOPLE TO WORK TOGETHER EFFECTIVELY

The Values-Based Workplace

5

Shared Values: The Eight Basic People Values

To educate a man in mind and not in morals is to educate
a menace to society.

Theodore Roosevelt

Since a lot of readers will not have read this book's preface, the central point is worth repeating here: the eight Shared Values—which form the bedrock for achieving lasting change—are derived from the analysis of 17 million surveys in which workers reported what they wanted in order to feel satisfied and rewarded by their work experience.

Some who read the list of eight values respond with a reaction like, "What's so special—I already live by most of these values," or "My company operates by these values."

The remark is generally only half true: they are values we're all familiar with, values we *intend* to live by. But just as we intend to obey the speed limit on the freeway, what happens when everybody around us is speeding? Of course we end up going with the flow and driving at the same speed as everybody else.

The strength of Shared Values within the organization lies in getting everyone up to speed together. When that happens, the values become not abstract goals we acknowledge as worthy, but practical, real-world principles and standards that we act on and are judged by every day.

These eight values are as fundamental to honorable living in the business community worldwide as are the Judeo-Christian Ten Commandments. The eight People Values are:

Truth: Treat others with uncompromising truth.
Trust: Lavish trust on your associates.
Mentoring: Mentor unselfishly.
Openness: Be receptive to new ideas regardless of their
                 origin.
Risk-taking: Take personal risks for the good of the
                 organization.
Giving credit: Give credit where it's due.
Honesty: Be honest in all dealings; do not touch
                 dishonest dollars.
Caring: Put the interests of others before your own.

1. TREAT OTHERS WITH UNCOMPROMISING TRUTH

Minneapolis-based Ault, Inc., a manufacturing firm with 165 employees, was in the process of obtaining from its bank a large increase in its line of credit. The new funds were crucial for sustaining operations at a time when the computer industry that Ault served was going through a painful period of slow sales and declining margins.

Ault had already filed application papers with the bank and received verbal assurances of approval. At that point the company got some bad news: one of its largest clients had just filed for bankruptcy.

The information given to the bank was no longer accurate.

Ault would not be receiving the large sum owed it by the customer, and some of the projected annual revenues shown on the forward P&L could no longer be expected. These changes would adversely affect both cash flow and receivables over the ensuing nine months.

The Ault executives faced a dilemma. If they shared the information with the bank and revised the financials, the company's position would not look nearly so strong and they would almost certainly be denied the increase to their line of credit. Without these new funds, the company would have to lay off valuable employees, who might not still be available when the computer economy strengthened, leaving the company less competitive.

At the same time, Ault management was confident that news about the bankruptcy would almost certainly not come to the bank's attention for months. And it wasn't even a question of lying: Ault could simply "play dumb," wait until the line of credit had been increased and the funds committed, and then let the bank know of the changed situation.

The good of the company seemed to demand playing it this way. Who would ever know?

Despite the strong temptation, Ault had embraced the Shared Values approach, and the Ault managers knew from the first that their corporate values required telling the truth.

The bank's reaction was not what the company feared. One Ault official later remembered, "By taking the risk of sharing the information with the bank and telling the truth immediately, we gained their respect." Indeed, the bankers were so deeply impressed by the company's straightforward truthfulness that the relationship was strengthened.

Truth-telling is so much a part of our fundamental value system that we assume it's a given. Closer examination of our own standards and behavior, and the standards and behavior of our children, our peers, and our bosses, reminds us that, as individuals and a society, our actions fall a good deal short of our standards, on a daily basis.

A fundamental premise of the Shared Values approach calls for establishing a standard of *uncompromising* truth—uncompromising in the sense of "unwilling to make concessions." That is, the straight truth, without shading or reservation, avoiding half-truths, not hiding behind the dodge of withholding. This is tough stuff to accept on faith, and yet the alternative is to be like the embezzling employee who never takes a vacation for fear of the surprise audit.

We list this as the first of the eight values because today's world seems to create so much uncertainty about what the truth is. Media reports, statements by politicians, press releases by businesses, even announcements from some less-than-ethical scientists and pharmaceutical companies so often today deal in white lies, half-truths, evasions, and innuendoes that the public—all of us—is suspicious and skeptical about much of what we read and hear.

Which makes the adherence to truth-telling all the greater a challenge, and all the greater a need.

Through years of experience in explaining and defining these principles in both private and public institutions, we've come to see that merely stating the values isn't enough. No matter how good the intentions, people need specific techniques to help them make each value a part of their behavior. Over time we discovered that the route to making this happen lay in providing specific guidelines—a set of practices to follow while you make the value become part of your practiced, accepted everyday behavior.

For truth, the guidelines are these:

Treat Others with Uncompromising Truth

"The 24-Hour Rule": Discuss the truth with the other person within 24 hours.

Despite the best of intentions, even the most saintly among us—on the spur of the moment, under the all-too-familiar pressures in our business and our personal lives, under the rush of events—sometimes blurts out a statement that falls short of our noble intentions of being truthful. This may occur because at the moment we are too emotional, or need time to reflect, or aren't prepared to handle the confrontation that the truth might bring.

This guideline sets a practical and realistic way to overcome the hurdle. It says that whenever you recognize you have not been entirely truthful with another person, you must *within 24 hours* return to the person and set the record straight.

If you believe someone hasn't been truthful with you, and they don't take action to set the record straight, the burden is on your shoulders: you need to get to them *within 24 hours*.

Often, correcting an untruth is even more important for you than for the other person.

"The Joan Rivers Rule": When telling the truth may prove awkward, painful, or embarrassing for you or for another person, obtain their permission first.

We refer to this as the Joan Rivers Rule because the practice makes use of a line similar to her well-known "Can we talk?" For Joan Rivers, the comedienne, it brought a laugh; for us, it gets the other person's attention and brings the opportunity for them to express a willingness to hear what you have to say.

You do this by asking, *"Is this a good time to talk?"* In a company that is putting Shared Values into practice, this specific question becomes reserved for the particular situation when one person wants to share an uncomfortable truth with another. So that the phrase will always be understood in this context, it should not be used for any other purpose.

The question is a formality, but one that demands acceptance and acquiescence: the other person has the right to say "No, not now." But saying "No" carries its own requirement: he must be willing to set a time when he will be prepared to talk.

What if the person won't set a time? At that point, the two people need to get someone else—some neutral third party—to step in. A peer, perhaps, or the boss, or the group or team leader.

If you are the person's manager, avoid making this request at the end of the day; don't leave the other person to worry overnight about what you want to talk about. Even worse is "Can we talk first thing Monday morning?"—which would leave the conversation hanging over a weekend.

Present the truth in a non-threatening way.

Just as the definition of quality depends on what the *customer* considers quality, so your evaluation of a non-threatening conversation depends on whether the other person considers it threatening. You need to watch for cues on how they are reacting, whether they are getting upset by your remarks.

Some will find this manipulative, but a technique that helps in this type of situation is what has been called "pacing," meaning to match the other person's body language—sitting as they are sitting, folding your arms if theirs are folded, averting your eyes if they avert theirs. These cues will speak for themselves. And it can be argued that setting another person at ease is never manipulative.

Talk straight, without hurting the other person's feelings, and be certain not to apologize for telling the truth.

Use language that is simple, understandable, non-apologetic, and non-personal. Not "You always do that" or "You never come to work on time."

When making a request, giving instructions, or remarking about behavior or performance of another person, avoid the word "you" altogether. Replace "You never come to work on time" with something like "The rest of us all count on everybody being at work on time."

Telling the truth is not a license to be hurtful. It does not give you permission to say "That's the ugliest shirt I ever saw" or "A college graduate shouldn't have the handwriting of a five-year-old."

But telling the full truth does put you under the obligation of sharing information even when it may be embarrassing or painful. After obtaining the person's permission in the usual way, it's acceptable to find a tactful way of saying what needs to be said. Some examples:

My request concerns how coming in late affects the others in the group. And the lateness is beginning to impact the quality of the assignments you're receiving. People are connecting coming in late with wondering whether you'll get an assignment done on time.

Or, *Tom, it seems to me that in business conversations, it's very important to stick to the subject that everyone else is trying to talk about. There was a team meeting the other day; as we were leaving, someone said, 'We wouldn't have been able to get so much done if Tom had been here because he would have kept changing the subject.' As much as that hurts, the problem of not sticking to the subject is having an impact on others. I wanted to be up front on this.*

Make a request, not a complaint.

Tell the other person how you would like things to be.

Not: *"You ship orders late much too often."*

Instead: *"I want your schedule for shipping times to be accurate."* To which you might add: *"If you need to, give me a bracket of times instead of a specific date. I need to be able to count on every team member being on top of this issue. Let me know how I can help."*

Not: *"Your meetings never finish on time."*

Instead: *"We all have tight schedules. We need to be able to count on meetings ending on time, so we don't keep other people waiting."*

2. LAVISH TRUST ON YOUR ASSOCIATES

A Midwest company where we have been installing the Shared Values process produces components for the electronics industry. Early in our work with them, the product marketing team was scheduled to go to an industry conference in Germany. The appearance was to be an important step in opening a European market for the company, and planning for the trip had gone on for months.

Despite all the meetings, schedules, lists, and plans, someone noticed at the last minute that the tools needed to set up the products and displays had not been shipped ahead with the other materials. As one of the team members later recalled, "It became obvious we would need to ship the tools by special courier or there wasn't going to be any show for us."

One of the senior team members, Dick Jordan (the names are fictitious), gave instructions to Kim Howard, the member representing the shipping department, to get the tools packed up and in the hands of Federal Express before the end of the day. Perhaps Kim didn't qualify as an "expert," but she had been on the shipping staff long enough to have accumulated some valuable experience. She said to Dick, "We've been working with this overseas carrier, and they can guarantee delivery when we need it. And their fees are significantly cheaper than FedEx international rates."

Dick told her, "This *has* to get there right away. FedEx can do it, and they have a reliable online tracking system. If we have to track the shipment once we get to Europe, it'll be mission impossible any other way."

At that point Kim was thoroughly disheartened. In a reserved tone, she told Dick, "I've been analyzing the rates that different shippers charge us for both foreign and domestic shipments. And I track how reliable they are in on-time delivery. Believe me, I can get the package there when you need it, for a lot less cost."

Kim understood she could benefit the organization by saving money on the shipping charge, and she was willing to put her reputation on the line toward this end. But it would be Dick's neck in the noose if the tools didn't arrive in time for the show. Still, he reversed his decision and put trust in Kim's experience and capability. By doing so, he lavished trust on Kim and expanded everyone's game in the process.

The trust was well placed. The shipment arrived when promised and as Dick later reported, "The company did save a great deal of money."

If you had been in Dick's shoes, what would you have done? Trust represents a philosophy about the confidence we place in others. At the same time, it's a reflection of the confidence we have in ourselves.

This isn't the kind of trust that involves a person's character or integrity, but a belief in the Shared Value of responsibility, commitment, and full communications. This is what forms our definition of trust.

When we choose not to trust our associates, we undermine their confidence, deny them learning experiences, and lose the benefit of the contributions they could be making. Worse, we poison the atmosphere of the workplace.

Years ago co-author Bill Simon served aboard a Navy destroyer under two different Commanding Officers. One was an experienced, competent, and confident ship handler; the other had been given two weeks of ship-handling instruction before taking command, and was in constant fear of misjudgment. The first allowed his young officers to "take the conn"—giving commands to the helmsman and engine room—in all challenging situations such as mooring to a buoy, docking to a pier, or steaming in dangerously close quarters to another ship during underway refueling.

And what about the other captain? Just what you would expect: with little trust in himself, he had little trust in anyone else. The junior officers got to do no more than stand around and watch while the captain, in a sweat and panic, gave all the orders himself.

What's involved here turns on more than just giving responsibility, which is an action; lavishing trust is a value, a spiritual idea that people can embrace. However, it's a very personal value—one person's level of comfort in handling trust is different from the next person's. To lavish trust, we must continually push the outer limits of our envelope, enabling those around us to truly play at the top of their game, a course that allows us to play at the top of ours.

Trusting people doesn't mean they will always succeed. Inevitably, sometimes we give an assignment, make a request, or share a responsibility, and the person doesn't come through. Even with a great deal of experience in placing trust, these can be tough calls, and each situation begins the process again. But even when the person fails, they will have the opportunity to learn from the failure.

Let's recognize that there are two forms of trust in business. One deals with honesty, which we deal with as another of the eight values. The other deals with whether the people around us will do their assigned jobs correctly. We need to recognize that underlying this form of trust is an issue of *communication*. Think about it: if we have not communicated the job assignment clearly, and been clearly understood, the other person will almost certainly not complete the job the way we intended.

In the earlier example, suppose Dick had told Kim, "Get the tools into Federal Express today." FedEx has almost become a generic term, like Xerox, Coke, or Kleenex. So what does the statement really mean? "Today" is clear enough, but the rest could mean "Send it by express carrier" or "Send it by the fastest delivery method" or "Send it by FedEx, and use your own judgment about which level of FedEx service to use."

Most of the details Kim needed to satisfy Dick are missing from this instruction. Kim's marching orders for shipping had always been, "Use the lowest-cost shipping that will get the item delivered on time." To Dick, the instruction to use FedEx meant, "This is urgent, use the fastest, most reliable delivery possible." Kim had been told to use FedEx, but not given any of the explanation of what was really needed—so one of the slower FedEx services would seem entirely justified, and could save shipping costs.

And what would Dick's reaction have been? In his anger when the tools didn't arrive on time, he would almost certainly blame Kim and conclude she couldn't be trusted. So an innocent misunderstanding might well result in mistrust, hurt feelings, frustration on both sides, and a deterioration of confidence in their relationship.

Too often we draw conclusions about trust, when the problem is really a lack of good communication. That's why our guidelines for trust offer ideas for improving communications—in two directions, both giving and receiving.

Guidelines for Giving

Give the other person a clear picture of what you want accomplished.

Describe the goal; if needed, be specific about how the task is to be performed (but *without* micromanaging). When possible, provide an example, a sample, or a model—for instance, a copy of a report you consider thorough, complete, and well written that you would like the other person to use as a standard.

Get agreement on completion time and what the finished job will look like.

Be specific about when you want or need the work to be finished. If you need it done quickly, find out what other projects the person may need to postpone in order to give priority to your task; is your task really more important than the things that would get postponed?

Do you want a verbal report or a written one? A working model of the new product or just a mock-up? A cash-flow projection for one year or for the lifetime of the product? Unless you take the time to be specific about what the finished job will look like, count on being disappointed.

(There's also a side benefit: by thinking through your request, you give yourself an opportunity to recognize challenges, problems, and loose ends in advance.)

Do everything you can to support the other person.

Sometimes in our trust of others, we delegate without supporting; that won't do. Once having given the assignment, you have to be willing to do everything you can to support the person—providing information, tools, resources—whatever she will need to carry out the task.

Be willing to trust the other person to complete the task, even if you see them encountering challenges.

This is the toughest part. You need to consider whether the assignment is appropriate to the knowledge, skills, and position of the other person, and avoid asking someone to accept an assignment she is not qualified to handle—unless you are intentionally challenging her to stretch (in which case, pay special attention to the guidelines here).

Trusting another person to complete the task becomes especially important when you give an assignment to someone with whom you have a long-term functional relationship; in that situation, trust means allowing them the leeway to complete the task even if you see them getting into trouble. Allow them to do the project *their* way. If they do it your way, they lose the opportunity for a confidence-building experience. And if things don't turn out well, they have an easy out: "I did it your way."

Guidelines for Receiving

When you're on the *receiving* end of an assignment, here are the guidelines that will let you safely accept someone placing trust in you:

You must be willing to agree to the task.

Have you been given the opportunity to agree without duress to handle the task? If you are unable to refuse the assignment, then you're acting under duress, and the assignment is not being made on a basis of trust but on a classic basis of delegation and control. As you start practicing this new trust model, you'll notice a big difference between the new model and the traditional delegation approach. People and groups ought to have the freedom to say "No."

That's important enough to be worth repeating: people and groups must be able to say "No." If they can't say "No," then they are only members of a chain gang, not self-determining individuals or groups, not willing partners.

Restate the request in your own words.

To be certain you have understood the assignment correctly, restate to the other person—the person putting trust in you— your perception of what the finished job will look like and the completion date—repeating in your own words the instructions that gave you his or her trust.

Get the trust-giver to acknowledge that you have stated the assignment correctly. This closes the loop.

Be willing to ask for help when you need it.

In most organizations, asking for help is a sign of weakness; in a Shared Values environment, it's a sign of strength and intelligence. If the people around you are not asking for appropriate help, you should be concerned.

3. MENTOR UNSELFISHLY

A worker on the production line at one of our client companies—among the largest glass fabrication firms in the United States—got into a heated dispute with his lead person: the worker had rejected a finished product, and the lead, who was new, was insisting the problem was minor and the product should be released for shipping.

In fact, the glass product itself was perfect; the only defect lay with the company's own colorful logo, which had become stretched and distorted while it was being applied. The worker was unwilling to let the product go out with a less-than-perfect logo, while the lead, answerable for production rates and completing work on schedule, insisted, "We're not selling the logo—the product is good and it goes."

The supervisor was called, and listened to each man explain his position. This was tricky; it would demand a good deal of tact. He had to make sure the worker would maintain his respect for the lead, and had to find a way of guiding the two toward reaching their own understanding.

Asking the lead to join him out of everyone else's hearing, the supervisor dealt with the situation as an opportunity to mentor, helping the lead grow into his new position. He carefully explained the importance of the logo, which reflects on the reputation of the company. Logos are so important, he pointed out, that major companies spend a million dollars and more on their design. Receiving product with a defective logo is like picking up a new car from the dealer and discovering it has a dent—the car runs just as well, it's just as comfortable, gets just as good mileage . . . but it's not perfect.

It seemed as if the lead was beginning to understand, yet he still raised an objection about the hit they would take on production rate if the product didn't ship.

The supervisor replied, "It's been my experience that good production rates will come when *all* details are taken seriously." And he went on to add that a supervisor is always on display, every minute of the day; how you present yourself is very important. "An incorrect impression will do more damage to communication and performance than anything else."

But he offered reassurance, as well, acknowledging that the lead was acting from a motive of wanting to do the best possible job. They talked about why the worker was upset, and how the lead could handle the situation without losing face.

At a Shared Values training session not long after, the supervisor reported that his greatest satisfaction of the whole experience was in seeing how the lead himself had learned a lesson about mentoring, and began to take a much more active role in mentoring others and allowing himself to be mentored, as well. The lead had learned an important lesson: that everyone can teach us something.

In Greek legend, Mentor was a friend of Odysseus, and the man to whom that intrepid traveler entrusted the care and teaching of his son. (But it was Athena, the goddess of wisdom, who disguised herself as Mentor to lead the boy on a search for his father.) Since the 1700s, the word has come to mean "a wise and trusted counselor."

In common usage today, the term conjures up a picture of a gray-haired older person who takes a younger associate under his or her wing, sometimes advising on the handling of day-to-day issues, but more often helping to shape and steer the younger person's career by offering wise counsel and, when possible, helping to put the person in line for increased responsibilities and career advancement.

That represents what we might call the "grand" or "high-level" style of mentoring, a marvelous opportunity that every younger person can hope to find one day.

But there's also another form of mentoring, little recognized yet offering far more opportunity for *everyone* in business—and not just on the receiving end but on the giving end, as well. This involves a form of mentoring that deals with single, unrelated, incidental opportunities.

Mentoring is like a sharp stick—it can be used constructively, or can inflict considerable harm. When an organization has effectively adopted a Shared Values philosophy, everyone learns how to give mentoring positively, and everyone becomes open to accepting it from others—no matter what the other person's position in the organization might be.

In using the guidelines that follow, remember they apply not just to long-term mentoring relationships, but even more to the workaday, onetime opportunities.

Guidelines for Giving

Get permission to mentor the other person.

Just as with giving trust, you need to ask for and receive the other person's permission before starting to mentor.

The recipient must have the right to say "No" or "Not now." In the same way that someone struggling with a crossword puzzle may not want your help because she's still hoping to finish it on her own, a person you're offering to mentor may not want help because she's still hoping to resolve the problem herself.

Whatever the reason, respect the person's request to be mentored at a later time.

Share knowledge, skill, and experience in a friendly way.

For an example of how *not* to mentor, listen to the way most parents give guidance to their children: it often comes out sounding at best critical, if not downright demeaning.

To have your mentoring appreciated and accepted, simply consider the situation from the other person's perspective, and couch your mentoring in language he or she will be able to accept.

For example, *"You may already know this, but I have a suggestion that might help. I've found that. . . ."*

Or, *"You may have come on this yourself, but something I've found is. . . ."*

Put the other person's interest before your own.

Remember that even if people give permission, your advice could offend. Be sure to consider their interest first, before your own. If it's clear it's out of the goodness of your heart, they will probably sense your intent.

Avoid grandstanding; let the other person retain his or her dignity.

Guidelines for Receiving

When someone wants to mentor us, we're often up against the wall on a deadline, or hurrying to get to a meeting on time, or in some other way feeling the pressures of the working world. An offer of mentoring rarely seems to come when the world around us is calm and peaceful.

And that's just one of the reasons why it can take a Solomon to accept mentoring. Being mentored is rarely easy. These guidelines will help:

Give permission to be mentored.

When someone offers you mentoring, learn to accept with good grace—the advice or suggestion will often be something that gives you the benefit of someone else's knowledge or experience, or that helps you work more effectively or interact better with your peers, subordinates, or managers. Most of the time, the information will be valuable and worth listening to.

As with accepting trust, you don't have to accept mentoring at the moment it's offered. If it's inconvenient at that time, or if you need to get yourself into the right frame of mind to accept mentoring from this person, your answer should not be "No," but "Later." Find a time that's convenient for both of you.

Remain willing to be mentored by everyone in the organization.

In most organizations, mentoring only goes downward—a more senior person offers mentoring to someone younger or newer on the job.

This pattern seems to be based on an assumption that those who are younger, less experienced, less senior, or in a different part of the organization have little of value to offer to anyone but their subordinates (*if* they have subordinates).

But almost everyone you know could tell stories about times when they were given a great idea or personal suggestion by a shipping clerk, a stranger on the telephone, a bag-boy at the supermarket . . . or by a child (we even have an expression for this one: "Out of the mouths of babes . . .").

So mentoring in the organization needs to go in all directions— not just down, but sideways and upward as well. Which means that we as individuals need to be open to receiving mentoring, no matter who wants to offer it to us.

Mentoring can be a great gift we give to others, and a great gift we receive.

Paul Horgen, CEO of the 200-employee IBM Mid-America Employees Federal Credit Union, has a favorite mentoring story from his organization:

"We had a woman who was having trouble with a high-powered computer program, a very complex spreadsheet for bank use. She got on the phone, called someone she knew in the accounting department and said, 'Will you mentor me with this program?' 'Sure, I'd be glad to; I'll come down right now.' He went down and in a few minutes she had learned what she needed."

Horgen adds, "Before Shared Values, almost anybody in that situation would have thought, 'I'm not going to tell anybody what I don't know because they'd tell others and the first thing you know, I'm going to have a reputation for being a dummy.' When you get people working together instead of worrying about what someone is going to think of them if they ask a question, that's powerful. And when you get that happening 100 or 200 times a day, you're talking *significant* improvements in productivity.

"We now have people saying, 'This is a great place to work, I like working here.' People in this organization will tell you, 'I don't ever want to work in any company unless they've implemented Shared Values.' "

4. BE RECEPTIVE TO NEW IDEAS REGARDLESS OF THEIR ORIGIN

Gabbert's Furniture is a high-end retail chain with headquarters in Minneapolis and outlets in Dallas and Ft. Worth, 650 employees, and over $100 million in annual revenues, and is considered by many to be one of the leading home furnishings retailers in the United States. The company has a reputation for hard-working people who excel in serving customers, and who have a sense of pride and accomplishment. Yet its people had become very disenchanted with how the organization helped them do their job. According to CEO Jim Gabbert, "People felt an enormous frustration that the hierarchy would get in the way—good ideas would be stonewalled by a manager and never make it through the approval process."

There had been some vague awareness in management

that you could be getting valuable ideas from your people, Gabbert says, but the management team had always thought, "We just need a better process, we just need to tell people that they must hold us accountable for listening to their ideas. So we'd go out and announce, 'We want your suggestions, we're not going to let your ideas sit on desks any more. And this time we really mean it.' "

Sometime after a major remodeling of the Minneapolis store, in which the accessories department (vases, pillows, mirrors, etc.) had been relocated, the salespeople in the department realized they should have been located on the next floor up, near the lamps.

Only a couple of years earlier, it would have been inconceivable for store clerks to suggest moving their department, and inconceivable that management would consider undoing a decision they had so recently made and spent money on. But in the interim, the company had instituted a Shared Values philosophy. As Jim Gabbert describes it, "Hierarchy was designed in the industrial age for workers who punched out the same thing time after time. That doesn't fit in the kind of environment most companies find themselves in today. The process we've gone through has allowed us to be who we are instead of pretending to be something else."

For Gabbert's Furniture, "who we are" now describes an organization where people are receptive to new ideas, a place where salespeople can indeed make a suggestion that will be taken seriously and, if appropriate, acted on.

The accessories department was indeed moved as the salespeople had suggested. With that distraction and several others at about the same time, Jim Gabbert had figured the company would be doing very well to have a flat year; in fact, he says, "We achieved record sales and record profits."

The guidelines that form a cornerstone of the way we explain the eight Shared Values grew out of a suggestion from a most unexpected source. After a class about truth for employees of a restaurant chain in our early days of teaching the values, one student, a woman who waited on tables in the restaurant, came up and said, "I know what the truth is, but I'm not sure everyone else understands it the same way I do. Can you put together some steps we can follow?"

If we had treated the suggestion with an attitude of "We're the experts, you just stick to serving the food," we would, of course, have been violating our own principle. In fact, the idea was so on target that it didn't take any coaxing or convincing; the guidelines you find in this chapter are the result.

But accepting new ideas regardless of the source turns out to be one of the most difficult principles for people to grasp and put into practice. Yet stories of brilliant outcomes abound, stories with endings as happy as a G-rated movie.

Management at a soft-drink bottling plant had committed to the purchase of a new processing machine. One problem: the new machine was so huge that a whole addition to the factory would need to be built just to house it.

Or maybe not. It was pointed out that by moving a couple of machines and rebuilding one section of roof, the machine could easily fit into the existing building. And what's more, with the repositioning, a single employee could double up to run two machines simultaneously.

The plant didn't have to fork over a fat consultant's fee to get this suggestion: it came, gratis, from one of its own production-line workers. Because management was willing to listen to a worker, the plant saved more than a hundred thousand dollars.

After employees at the restaurant chain Mitzel American Kitchens (see Chapter 1) gained courage to express their ideas as a result of the introduction to the Shared Values philosophy, a cook came forward with something he had been thinking about for a long time but had never mentioned before because he was sure no one would listen. It was a simple idea for keeping pies fresh longer. Because Mitzel's does such a huge business in pies, the little suggestion resulted in a 2 percent improvement to the company's bottom line.

Employees at Minneapolis-based ME International, a one-time subsidiary of Armco Inc. and half-owned by Stelco, Inc., decided after introducing the philosophy of Shared Values that getting rid of the corporate parking places by the plant entrance would be an effective way of signaling change. President John Oertel was reluctant because he didn't want to offend his managers but also recognized that rank-and-file employees may have begrudged executives this convenience. When the change was made, it took less than half an hour for word to spread through all the company plants in Duluth and St. Cloud, Minnesota, and even as far away as remote Ishpiming on the upper peninsula of Michigan.

Marcon Coatings, Inc., a company that produces coated glass sheets used in thermal windows, had developed a rigid, quality-driven procedure to be used whenever a production line changed from one type of coating to another. The new setup was tested by sending a piece of glass through the machines, requiring three test passes before stability could be assured. And at about eight minutes a pass, a lot of production time would get eaten up. But experience had shown this was the only way to be certain of quality.

A Marcon foreman described what happened: "One day we had a new night-shift operator who showed up and announced he had a way to dial in a coating test in a *single* pass. We all knew better, but we were trying to live our Shared Values of being open and receptive to a new idea regardless of the origin. So as a courtesy to the earnest young man, we let him show us how to set up the coater his way.

"It *worked!* Several helpings of 'crow' were dished out that day to some experienced Marcon managers who had learned a valuable lesson. This and several other procedures from this creative young man's mind are now the standard in today's operations."

Be Receptive to New Ideas, Regardless of Their Origin

Listen with an open mind, putting aside traditional thoughts and biases.

Easy to say, not so easy to do—witness all the talk in business books, corporate communications meetings, and annual reports about the need to change. If it weren't part of the human condition to be resistant to change, there wouldn't be such a compelling pressure to talk about it all the time.

Being receptive to new ideas starts with being a good, attentive, open listener. And even ideas that seem worthless on first hearing may turn out to have some kernels of value, if we're patient enough to think about them, weigh them, examine them. Few suggestions we receive are all or nothing, now or never. Highly successful people are almost always people who have found in many different places, from many different people, the ideas that made them successful; being open to new ideas is an almost universal strength of the truly successful. Most ideas come to us half-baked; it's up to us to cook them till they're done.

Remember that the more humble a person's position in the organizational structure, the more difficult it will be for you to listen with a truly open mind. One restaurant manager pointed out what he called the best food expert in his restaurant. It wasn't the chef, not even a waiter, but a man who cleaned the tables. "He picks up the dishes, and he's observant," the manager said. "If I want to know what's working and what's not, all I have to do is ask him." When you want a realistic viewpoint, are you overlooking people who might have the best answers?

Listen to ideas in a non-judgmental way regardless of the origin.

When presented with a new idea, a lot of people play devil's advocate, claiming they're testing the idea—when what they're really doing is throwing cold water on it and making the person feel foolish.

This discourages not just the person offering the idea, but everyone else who observes or hears about the incident. Remember that it takes a lot of courage to offer a suggestion—especially to one's boss or someone even higher in the organization. (Some of our client companies have created lapel buttons that read, "No devil's advocates.")

Restate the idea in your own words for clarification.

By now the notion of restating what you've heard is familiar. It applies here, as well: make sure you've correctly grasped what the other person is saying by repeating your understanding of what you've heard and allowing them to correct you if you've misunderstood or gotten some of the details wrong. Don't simplify; state the idea as fully as you were told it.

Ask questions for the success of the idea.

Asking questions not only clarifies the suggestion, but also validates the other person.

"Would it work on the night shift as well?"

"Would it require more people, or do you think the existing staff could handle it?"

"How would it work with _____?"

Note that this must not be used as a way of poking holes in the person's idea; your purpose should be to try to make the idea work. Let the idea-giver see you are supportive and trying to figure out how it might work.

"Houston, we have a problem." During the Apollo 13 moon mission, an oxygen tank ruptured, leaving the spacecraft seriously damaged and the lives of the crew gravely threatened. Working with the meager data available, crews at contractor plants all over the country set to work simulating the condition and soon knew there was not enough air, water, or electricity in the spacecraft to sustain the three men.

But how about if they crowded into the lunar lander? The simulations said Yes. With ground crews devising ingenious solutions to a continuous series of problems, the spacecraft was brought safely back to earth. The happy outcome had only been possible because hundreds of engineers, scientists, technicians, and astronauts were simultaneously asking vital questions and finding essential answers; there were no devil's advocates smugly sitting back and poking holes in each new idea.

Agree to provide a follow-up.

Once you've been given a suggestion or idea, how do you end the conversation—"Thanks, I'll think about it"? "Joe came up with something like that two years ago, and it didn't work then"?

Your way of responding will color people's views about whether to bring you ideas in the future. Even if the suggestion is impractical, too expensive, or in some other way unworkable, you need to treat it with the degree of respect that will encourage the future flow of ideas.

The moment of receiving a suggestion is not the time to turn it down, no matter how unwieldy or unmanageable. Instead, give the person a time and place when you'll respond.

And then be sure you keep to your schedule. Failing to respond in a timely fashion is one of the great sins of managing and working with others.

5. TAKE PERSONAL RISKS FOR THE GOOD OF THE ORGANIZATION

One person who thinks he is right makes a majority.
George Washington

In the early 1960s, IBM launched a project to develop a high-end computer, code-named Stretch. The machine that reached the market achieved no more than 70 percent of the original design goals and failed to meet customer needs. Thomas Watson, Jr., who was then running the company, canceled Stretch even though it meant a loss of some $20 million, and sent out a message for the vice president responsible to come see him.

The VP called his wife and said, "Pack me an overnight bag, I have to fly in to see Watson, and when I get back I won't have a job."

He arrived at corporate headquarters the next morning, ashen but determined to remain calm no matter what. When he was shown into the office, Watson gestured to some manuals on the desk and said, "I've put together some materials for your next project."

"I thought you called me in to fire me," the surprised VP said.

"Fire you?! Why would I fire you?" Watson answered. "I just spent $20 million on your education."

People who never fail are people who never take risks. Organizations and work groups that never fail are those that never take risks. Thomas Watson knew that.

CEOs frequently wonder, "How much leeway should I give?" Tactical, highly bureaucratic organizations drive the life out of their people by not giving them room to make decisions and room to make mistakes.

One of the fundamental human needs is the need to express creativity, not necessarily in an artistic sense as a painter, writer, musician, or filmmaker, but in the sense of making individual, personal contributions that move a work group, a department, an organization, or a family toward its goals. We all need to become better at stepping forward with new ideas, even ideas we suspect will be unpopular.

It's worth repeating that advances in civilization and science are not achieved by organizations but by individuals, or a small number of individuals acting as an inspired team.

Some organizations are inherently comfortable with risk-taking. The Boeing Aircraft executives and Board of Directors in the late 1960s made a decision to build jet transport airplanes, knowing that they were putting the entire company at risk. (As the story goes, the design concept for the Boeing 747 was first conceived on a napkin in a bar in Asia.) At General Electric, CEO Jack Welch and CFO Dennis Dammerman have approved a billion-dollar development program for a next-generation jet engine, despite their understanding that the endeavor will not begin to generate income for the company until about the year 2020. And it was not uncommon for Konosuke Matsushita, the founder of the Japanese industrial giant Matsushita Electric Co., to plan projects with a span of 25 years' duration; under his guidance from 1918 to 1973, the company's revenues grew some 4,000-fold.

Some individuals, too, find risk-taking not only acceptable but appealing. Steve Jobs put Apple Computer on the line when he insisted on developing the Lisa, against the counsel of many Apple executives, managers, and technologists. The Lisa failed . . . but made the Macintosh possible.

An old *New Yorker* cartoon shows a businessman standing at his closet to pick out his clothing for the day. The suits lined up on hangers are all the same color, all identical. On the tie rack are two ties, both reps, but with the stripes slanted in opposite directions. His choice each day is limited to which direction the stripes on his tie will go. We laugh because we intuitively understand that people want more choices—in clothing, in life, and at work.

At the same time, risk-taking is a highly individualized issue. Everybody tests their own level, and one person's comfortable risk is beyond another's threshold. Chuck Yeager, the test pilot who was the first man to break the sound barrier, proving that an airplane would indeed not break into small pieces on encountering the supposed barrier at the speed of sound, once appeared on the *Tonight Show with Johnny Carson.* When Carson said, "You've taken a lot of risks," Yeager replied, "You might think so, but I never put myself in a situation I didn't think I could get out of."

Here was a man willing to attempt something that many aeronautical engineers of the time thought would be disastrous, yet who proceeded with enough care that he evaluated the situation as offering an acceptable level of risk.

Perhaps we should all put a statue of Chuck Yeager in the lobbies of our buildings or on our desks, as a reminder to continually reshape our organizations so that we encourage an appropriate, acceptable level of risk-taking. Let's work toward establishing a new attitude, one that says "People don't get punished for failing, they get punished for not trying."

A book some years ago, *The Vital Difference,* highlighted companies that actually celebrate failures—for example, by throwing a beer blast to announce a project that had collapsed or been canceled. Too extreme? For most companies, without doubt. But it certainly gets the message across that taking risk is smiled upon and applauded, even *expected.*

And there's the oft-told story of the 3M researcher who presented a report on the failure of his attempt to develop a new glue. The best he had been able to achieve was a glue that would stick two pieces of paper together, but wouldn't form a permanent bond. Listening to this report of a failure was a man who sang in his church choir, and was constantly having problems because the little scraps of paper he used as place-markers were forever falling out of his hymnal. That non-permanent glue sounded like it might be just what he needed. The result, of course, was the Post-it Note, which has been immensely popular since its introduction.

Take Personal Risks for the Organization

Be willing to share ideas, even if they might be unpopular.

So often we find the fear of the unknown much more terrifying than the reality we face in stepping forward with a new idea. In the memorable book *Paradigm Pioneers,* and the business video of the same name, author Joel Barker suggests that many of us have had but never acted on ideas that later became today's latest hot product, today's newest industry. He talks about people with whiney voices who claim, "I had that idea years ago."

So why didn't they jump in? Barker's answer: "No guts!" Obviously it's a lot more complicated than that, but ultimately we need to create a work environment where speaking up to share an idea is possible.

Be willing to speak up even when you know you might be wrong.

Even though Steve Jobs was head of Apple when he took the risk of building the Lisa, precursor of the Macintosh, and Welch and Dammerman were the leaders of GE (and still are), when they decided to risk developing the new jet engine, you don't have to be head of an organization to speak up about a risky venture or a new way of doing something.

In fact, people find it easier to speak up and take risks when they are *not* the head of their organization—since they have less at stake, and since their risk-taking will require the endorsement and support of a boss or team leader and members.

Be willing to speak up or forge ahead when you believe you're right, even if your information will not be popular.

We learn early in life—on the school playground and elsewhere—the value of being popular. Those early lessons make us wary about speaking our minds or taking a stand if it might be unpopular with our co-workers or management.

That's a hurdle we need to overcome; but it's easier to overcome once we recognize that it is, indeed, a hurdle standing in our way.

Recognize and support others when they step forward with their ideas.

Risk-taking becomes a whole lot easier when others are willing to lend support and join us in our risky venture.

Like so many other things in life, this one works on a reciprocal basis: if we want the support of others, we have to be prepared to give them our support when they're in need of it. First decide if the project makes sense (see the next guideline); once you've determined that it does, give the same kind of Golden Rule support you yourself would like to receive.

Use common sense and make your actions appropriate to the situation.

Even though what we mean by "common sense" is rather vague at best, subject to wide individual interpretation, it's still the most reliable measure for determining the appropriateness of a particular risk.

The heroic act of personal risk-taking, when the act is for the good of the organization and not just for promoting a person's career, is an important addition to any Shared Values environment.

6. GIVE CREDIT WHERE IT'S DUE

The question is sometimes asked, "When should a man tell his wife he loves her?"—to which the best answer is: "Before some other man does." In the same way, you need to let your employees know they are appreciated, before some other boss gives them the appreciation they're not getting from you. How often do we say to the people around us that we need them? And yet, how effective that phrase can be to let people know how much we appreciate them.

The act of giving credit serves two functions: it's not only a way of acknowledging people who have done an especially good job, but also, less obviously, a way of upholding the standards of the organization. That second function, while highly important, doesn't get noticed much, yet the consequences of overlooking it are widely felt: organizations that don't respond when their standards are not met tacitly allow managers, employees, and outside influences to set lower standards.

So giving credit and adhering to standards are flip sides of the same coin. If you can give credit, you can also make requests that will lead people to uphold the standards of the organization.

The guidelines for giving credit thus also deal with the issue of maintaining standards, which helps create the context of our organizations.

Give Credit Where It's Due

Give praise that is appropriate and genuine.

Giving credit generally takes the form of giving praise; the danger when learning to give praise lies in turning yourself into a cheerleader. If this happens, your praise becomes less and less sincere, and less and less valuable.

Some think that giving praise leads people to have lower enthusiasm or energy. There's no evidence to support this notion. The evidence of experience clearly shows, though, that overpraising, when done to a degree that's perceived as not genuine, has about the same negative impact as underpraising. So—give plenty of praise, but only when it's truly earned.

Give credit and praise while the job is happening—on small jobs as well as large.

Make your praise timely—while the work is being done, or soon after it's finished. A thank-you, along with a recognition of the completed job, is often praise enough. Be careful to take notice of small jobs done well—they're much easier to overlook than major projects that command a lot of attention. Successful large jobs occur because of all the small efforts that support them.

Many people have been taught to "say something nice" before giving a difficult assignment or asking for something special. But the "something nice" too often takes the form of praise, and that creates a problem. It's like the little boy whose mother always masked bad-tasting medicine by mixing it into raspberry ice cream; he grows into a man who hates raspberry ice cream. If you always precede a request with praise, every time you give praise, people will freeze up, wary that you're about to ask for something.

So make your praise immediate, appropriate, and straightforward, and avoid using praise as a prelude to a request.

Instead of criticizing, use positive statements about what you want.

Everyone who supervises people routinely gives guidance and coaching. The traditional way of doing this has been in the form of criticism—which has an unfortunate effect: it comes across like you're telling the person how much smarter you think you are.

Even constructive criticism is one of the least effective ways of coaching—most of the time the person doesn't think the problem was his fault and doesn't want to accept the criticism, even when a compliment is thrown in. You run the risk of creating an unpleasant situation at the least, and at worst turning people hostile, uncooperative, and unwilling.

To communicate more effectively, avoid statements that sound like you're criticizing; instead, make a positive statement. Try to get in the habit of starting this kind of statement with words like "My request is . . ." or "Here's what I need . . ." because this language alerts the person to what's coming.

Another language tip: try using the words "less" or "more" in your statement: "My request is that I need more color" or "What I need is more of an effort to get to meetings on time."

People tend to be very honest when they sense you are collaborating rather than criticizing.

Give credit fairly to all, regardless of their role in the organization.

For example, if a team does a stellar job, most often everyone on the team deserves the praise, not just the leader.

In particular, it's easy to overlook good work by very junior people or people whose work seems routine. When is the last time anyone in your organization gave praise to a cleaning crew, a window washer, a cafeteria worker, or the people in the mail room?

Give timely and appropriate feedback where you see actions that are not up to individual or group standards.

Many people recognize feedback as an effective technique, one of the things a manager or supervisor regularly needs to do. But be careful: the way you give feedback can make it welcome and acceptable, or offensive and trouble making. One problem comes from giving feedback about something over which the other person has no control.

If you are going to give feedback, make it as a request—something the person can do something about. Telling a co-worker that the overheads he presented at the meeting were the ugliest you've ever seen may be the truth, but what have you gained?

The best kind of feedback asks people to look at what they've done, and then offers a clear picture of how they could have handled the situation better. Language you can use:

"Jack, if you were going to give yourself feedback, what would you have done differently if you had had more time?"

"Maria, how do you think that speech went?"

"Alan, if you had used the graphics department to prepare visuals for your presentation, what do you think they might have been able to improve?"

Treat feedback as a debriefing session, the same way a football team on Monday reviews films of Sunday's game to learn what they did right and what they did wrong. But above all, make sure the other person's dignity is kept intact.

And give the same kind of feedback after a success as you do after a failure.

7. BE HONEST IN ALL DEALINGS; DO NOT TOUCH DISHONEST DOLLARS

Some years ago, when co-author Bill Simon lived in Washington, D.C., he served as president of the Foggy Bottom Association, the neighborhood group for the residential community that was also home for George Washington University, the State Department, and the sometime-infamous Watergate office and condominium complex.

One of Bill's friends, real estate developer John Safer, had plans to build a new high-rise apartment building on 25th Street in the neighborhood. It turned out that the building could be larger, and considerably more profitable to the developer, if a way could be found to create additional parking. John's architect came up with an idea, but it required creating a new alley that would run alongside the building to a rear-entrance underground parking access. A good solution, with only one hitch: the new alley would be approved by the city authorities only if the association agreed.

The request placed Bill in an awkward position: to arbitrarily deny the request would be unfair, but to approve it might have the appearance of an inappropriate favor to a friend. So when the matter came up at the association meeting, Bill turned the chair over to the vice president. Then, speaking from the floor as a member, he proposed a possible compromise: the group could consider allowing the alley, if the developer were willing to turn a nearby vacant piece of public land into a children's playground. The idea was enthusiastically endorsed by the membership: the developer would have his alley, and the community would gain a playground.

But Bill's problems weren't over. In gratitude, the developer showed up at his door with a valuable bottle of 60-year-old brandy.

Not an easy gift to turn down. And it was being offered after the fact, after the decision had already been made—clearly not a bribe or inducement. What would you have done?

Bill said later, "I knew from the first moment that I couldn't accept it. Even though no one but John and I would have known, it was too easy to misconstrue. And if it ever had become public knowledge, there would have been no way I could prove that the gift hadn't been offered beforehand. So I kept a clear conscience, even though I've always wondered what that 60-year-old brandy tasted like."

A cover story in *USA Today* (April 4, 1997) announced, "48% of Workers Admit to Unethical or Illegal Acts." Even more alarming: the figure was based on a survey which only asked employees to list unethical acts due to pressure from long hours, sales quotas, job insecurity, personal debt, and the like; it did not tally unethical or illegal acts for reasons such as revenge, anger, jealousy, or greed. The acts admitted to included cutting corners on quality control (the most common on the list), as well as "cheating on an expense account, discriminating against co-workers, paying or accepting kickbacks, secretly forging signatures, trading sex for sales and looking the other way when environmental laws are violated," according to the article.

Doing what's right calls on you to ask these four questions:

Is it good for the individual or customer?
Is it good for the group, team, department, region?
Is it good for the company or organization?
Is it good for the community and the environment?

A similar list is used by Rotary clubs; for them as well as for a business organization, it's effective in giving people a filtering device when a tough situation occurs.

Honesty has many faces; one is reliable, consistent integrity. You build pride in an organization from the honesty and integrity of the individual people. And like most every other values-based issue, honesty and integrity start at the top. If the CEO or a general manager or plant manager, say, sends some craftspeople from the shop to rebuild the kitchen in his home, it takes very little time for word of that special treatment to spread throughout the company. And the principle of *What we allow, we teach* swings into effect with predictable force.

One of the early CEOs who brought the Shared Values philosophy into his company told a story on himself about honesty.

While visiting one of his stores, he wrote out a personal check for $100, placed it in a cash register, and took out five $20 bills. His operations officer immediately challenged him in front of the store manager and assistant manager. "You put a policy in place a month ago that employees could no longer cash personal checks."

"Wait a minute," the CEO answered with an edge, "I own this damn place."

"Sure," said the operations manager. "But if we expect employees to follow instructions like that, *we* need to follow them as well."

"Who would know?" the CEO demanded.

"We know, and so will the people who post the store checks. Your accounting staff will know. That makes 11 of us— even before people start spreading the story around."

The CEO heaved a sigh. But he went back to the register, took out his check, and returned the $100.

The other side of the coin is that honesty and integrity can't be mandated; they don't happen just because the CEO announces that it's company policy. Instead, each organization needs to culturally define what it means by these terms in a generic sense, and then translate them into specifics for each operation and activity of the company. And then everyone in the company needs to learn and embrace those definitions and specifics, until they're thoroughly comfortable with how to put them to work as part of everyday behavior. After that the tough part starts: consistently "walking our talk."

Honesty has a special place in the scale of values for most people. In their enlightening 1993 book, *Credibility,* authors James Kouzes and Barry Posner report the results of a study in which employees were asked to identify the characteristics they most wanted in their leaders and in their colleagues. The degree of difference in the two lists is fascinating. For example, people want their leaders to be forward-looking, but put this way down the list, item #11, of qualities for a colleague; people want their colleagues to be cooperative, but rank this low on the list of qualities for a leader.

Yet on one value, the respondents placed the same importance: honesty was, by a wide margin, considered the most important value for both leaders and colleagues. (And isn't this characteristic the most valuable as well in your marriage, friendships, family, and with your children?)

What does "treating customers with honesty and integrity" mean in your organization? To answer the question, the organization first has to agree on what levels of service quality it will provide; until everyone is clear on the standards, you have no basis for assessing whether you are meeting your obligation of honorably providing quality service.

In all of our research, when people are asked which of the organizations they deal with has the highest overall service quality, the top places on the list go to retailers of Japanese cars, in particular Lexus/Toyota and Acura/Honda. But GM's Saturn dealers now also rank near the top of the list. If even corporate gorillas in a slow-to-change field like the automotive industry can learn to set new standards of honesty in dealing with customers, you have every reason to expect success within your own organization, as well. (Incidentally, Saturn began its organizational development centered around a set of shared values; you can go into any Saturn dealership and ask for a copy.)

To define levels of customer service, look for a model—even outside your own industry. In bygone times, companies talked about aiming to be "the Cadillac of our industry." Cadillac has (hurray!) recently returned to being a quality car, but during the years when it fell short of the earlier standard, the metaphor fell out of the language (and, curiously, was not replaced).

These guidelines will help you get started in the right direction.

Be Honest in All Dealings; Do Not Touch Dishonest Dollars

Be aware that employees need to feel proud of the products and services their organization offers.

Every organization needs to recognize that to create a high level of honesty and integrity, its people must feel proud not only of the products and services, but also of the values that the organization lives by.

Honesty is too abstract a concept for people to discuss easily. So talk about it in terms of pride, which people can grasp and resonate to much more readily. Begin with a companywide discussion on what constitutes pride in the products and services. People want to be, are willing to be, prideful without becoming arrogant. The discussions lead individuals to express what for them will constitute pride in the products and in their work.

One successful approach: in group gatherings throughout the organization, ask "What would it take to imagine being enormously proud of what we do? What would the place look like, what would the work experience be like, what would our relationships with customers and colleagues be like?"

Define your own personal standards of honesty, and make them non-negotiable.

If some of the people in your organization lack a fully developed sense of honesty, you can explain the principles but you won't likely bring about a change. Honesty is one of those values that we learn as children, or never learn. Remember, you can't fix people (which is in part why our jails don't work).

You can, however, help people *sharpen* their sense of honesty. Every member of the organization, regardless of level or position, has an obligation to help create the conditions where honesty is recognized as a virtue, not a political liability.

One question to pose: "What is non-negotiable to you?" The *Wall Street Journal* published an article (1/2/97) listing countries where bribes are required for doing business. In order to get a large and lucrative contract in one of those countries, would you consider it acceptable to give a gift to an official? Would you support your company in doing business under these circumstances? Would you consider it acceptable for a company you own stock in?

(As a sidelight, the article goes on to point out that there appears to be a clear correlation between a country's corruption factor and its economic-development success rating. Business ethics seem to be distinctly related to economic viability; worldwide, the most ethical nations enjoy the highest standards of living.)

Suppose the government official who is head of a foreign nation's air ministry is coming to the United States to see a demonstration of your air traffic control system, and your company is paying for his trip. Is it acceptable to invite him to bring his wife along, also at company expense? Suppose you do this, he accepts, and you learn it's not his wife he means to bring, but another woman? Most American companies do not involve themselves in censuring employees over whom they live with; in fact, it would be against fair employment practices to do so. Then would your company be out of line to tell the air minister who he could and couldn't bring? (Hughes Aircraft actually faced this problem. If you're interested in its decision, ask us.)

The organization must help each individual to determine his or her own standards. But, in a way that's fascinating to observe, when a company adopts a Shared Values philosophy, employees perceive the organization as bringing *its* standards up to *their* level.

Clearly understand the organization's values and standards, and be prepared to support them.

Strong leaders historically embody the values and standards of their organization. When those individuals leave, the organization's standards sometimes walk out the door with them. The only way of avoiding this is to insure that the standards become part of the fiber of the organization.

Even in the absence of a companywide Shared Values philosophy, you can do your own part by making sure you understand the values and standards that define what the organization truly stands for.

Be prepared to take action if you see any touching of dishonest dollars.

In 1987, when co-author Rob Lebow was Director of Corporate Communications at Microsoft, he was in the process of selecting a new advertising agency and preparing for the launch of Microsoft's newest and most strategic product, Excel, designed to challenge the then-leader of the spreadsheet marketplace, 1-2-3, the flagship product of Lotus Corp. In the midst of this, he received a Federal Express envelope one morning with a most unexpected offer inside.

It was from an East Coast advertising agency, dangling the bait of some new employees who had just been recruited from another agency where they had been working on the Lotus account and so were "intimately acquainted with Lotus' . . . plans to deal with the introduction of Excel." Included in the envelope were round-trip tickets to Boston.

Rob remembers, "I didn't want to touch this with a ten-foot pole. I read it, sealed it back up, and had my assistant hand-carry it to the Microsoft lawyers."

What surprised him most, he now says, was not the offer but the reaction that followed: newspaper writers reporting and analyzing the incident, magazine articles praising him, a Harvard Business School case study exploring the ethical issues.

In the final analysis, his action had been automatic, requiring no thought or soul-searching. He never counseled with anyone at Microsoft about the incident and was never thanked by any of the higher-ups for his actions. But his staff gave him a standing ovation, and all the uproar made clear that a great many people considered his action noteworthy. There could only be one explanation: many feel a sense of uncertainty, either about their own level of honesty, or about the honesty of the people around them.

For most people, the question of honesty is easily answered. But here's the next question—and it's a really touchy one: are you prepared to take action if you see *someone else* behaving dishonestly at work?

And if you encounter such an act and report it—are you blowing the whistle for the good of the organization . . . or for your own personal advantage?

8. PUT THE INTERESTS OF OTHERS BEFORE YOUR OWN

Building heavy-duty dump truck boxes is one of the tough jobs at Crysteel Manufacturing. During the summer, when temperatures in the welding shop can hit over 100 degrees, the foreman arranged to give the men extra breaks.

Sometimes it's the little things in life that make all the difference: a highlight of each break was the chance to have a Popsicle. But one of the older work team members—we'll call him Brad—always stood alone and never shared in the Popsicle pleasure. Finally someone asked, and discovered he was a diabetic. The very next day a box of sugarless Popsicles was in the freezer with Brad's name on it.

Almost overnight, this simple act of consideration changed Brad from a loner to a person eagerly taking part in the group. Eventually he became a volunteer for overtime so the younger men could be home with their kids. A simple act of selfless behavior on the part of one person changed Brad's involvement with his work group—and the company has been drastically improved for it.

At the organization where this took place, the story has become a symbol of the power of Shared Values.

Management consultant Dr. Arynne Simon is occasionally asked to counsel a senior manager whose main complaint turns out to be that he or she feels lonely, remote from other people, disconnected. Most counselors advise people in that predicament to join some clubs, take some lessons, get involved in activities. Not Arynne; her advice is quite different: go volunteer at a shelter for the homeless, spend a few hours a week helping at a home for abused women or runaway children, or find some other need in the community that matches your interests and concerns, and give it your time on a regular basis. She reasons that joining a club or fitness center, or taking French lessons, simply focuses the person on his or her own needs and wants. But loneliness usually stems from a difficulty reaching out to others—something that many otherwise stable adults go through at one period or another in their lives. The antidote is not to be found in an activity that turns the attention further inward, but in something that involves opening outward by giving to others.

Sound advice . . . and not something just for members of the lonely heart's club. One of the driving universals about people is the compelling desire to affiliate. And that's something we achieve when we put others first.

People ask whether this is really an effective strategy for the business world. How can you put other people first, when they may not put you first? How can you put other people first, when every company and every individual seems to be driven by competitive motives? The answer is that putting others first allows people in the workplace to feel comfortable, safe, willing to participate, willing to take risks. It's the essence of heroic behavior.

Externally, the attitude leads your suppliers and customers to believe you'll deal with them honorably and with integrity. Customers will even pay more to buy from a company that treats them honestly, while your competitor companies end up wrangling for business on price alone.

One word of caution: some companies try to adopt an

approach of putting the interests of others first, find it doesn't bring short-term results, and move on to search for some other approach. The most common reason for this failure lies in their adopting the approach simply as one more program, a gimmick almost, rather than as a fundamental philosophy of the business. The moral: if you don't really believe in the underlying values, the approach will not help create long-term relationships. Many retail chains have tried the Nordstrom return policy of accepting merchandise back even if the customer doesn't have a receipt; but embracing this policy as a tactic and not as a shared value, the other chains have never had success, and have reverted back to a traditional, restrictive policy on returns—much to the chagrin of their customers.

Put the Interests of Others Before Your Own

Be willing to put the interests of others before your own.

The concept of the servant-leader dates back to ancient times. Nearly 1,000 years ago, the great Chinese philosopher Lao-tzu wrote, "To lead the people, walk behind them." The best leaders almost vanish: people do not notice their existence. When the best leader's work is done, the people say, "We did it ourselves."

In business as in our private life, we look back on our progress in the world and know that others have given us their shoulders to propel us up the steep cliffs of our journey. We, too, have a duty to do the same for others. No one gets there alone.

Root for another person's success, or the group's success, even when it appears there's nothing in it for you.

The poet John Donne wrote, "Never send to know for whom the bell tolls; it tolls for thee." Though he was writing about a sad event rather than a glad one, the principle still holds. When we root for the success of others, their every achievement gives us extra pleasure; we benefit, albeit remotely, from their good fortune.

Support others and yourself by acting upon the organization's values every day.

Wouldn't it be wonderful if everyone asked themselves before taking an action or making a decision, "What's in the best interests of my group, my organization?" In that world, we could pass decision making downward to the very people who make it all happen, instead of retaining control in isolation.

Examples abound of *not* putting the interests of others first. You only have to say "Exxon Valdez" for people to know immediately what you're talking about. Faced with having caused a disaster of epic proportions, the company dragged its feet about acknowledging responsibility, was slow to begin cleanup operations, and in the end did far less cleanup than the situation called for. To many people, the company name still brings to mind a crippled, leaking oil tanker and an organization that did not live up to its obligations to people and the environment.

In contrast, recall the Pepsi needle-hoax story at the beginning of this book. With a values-based environment in place, the company's first actions were aimed at consumer protection and safety, with no thought of what it might mean to reputation or revenues. Company representatives knew they had authority to take action without having to get permission or instructions from their bosses. They acted to protect public safety despite knowing that the claims of hypodermic syringes in soda cans couldn't possibly be true. And the outcome was public praise and *increased* revenues.

WRAP UP

How important are Shared Values? Every now and then you will be called upon to bet your company on them—making a decision about a recall, about releasing a product that will have to be either late or not thoroughly tested, about your relationship with a vendor, about changing sales goals, about publicly admitting some momentous error.

If the right values are in place, you and the others in your organization will have no difficulty making the right choices.

6

Business Values that Balance the People Values

Always aim for achievement and forget about success.

Helen Hayes

TWO SETS OF VALUES

It's our view that people share *two* sets of values, one dealing with how we treat people, the other with what we must do every day to be successful in the marketplace of our industry or service. These two sets—People Values, discussed in the previous chapter, and Business Values, which we explore here—are always in "dynamic tension." And they spring from different starting points: Business Values are crafted in a *competitive* environment, People Values in an *inclusive* environment where individuals interact with one another.

An organization's values are almost always carefully crafted and are usually taken seriously. But when used inappropriately, values can be wielded like a sharp stick. We sometimes half-jokingly discuss the disease of becoming a "Values Nazi." The telltale signs: a person who takes each phrase literally, who points fingers, who judgmentally calls attention to everyone's faults.

Values should nurture us, enrich us, not spell our confinement. If our values are not aspirational and embraced through informal "teachable moments," they will simply fall into disuse and be remembered only from those mostly ignored framed statements hanging in the corridors.

Soon after co-author Rob Lebow had taken over a 5,000-person sales division at Avon Cosmetics, an incident occurred that would later help shape his appreciation for Shared Values. An Avon representative phoned to complain about dishonesty in a nationwide sales contest of a new perfume: one of the competing district managers and several of her team leaders were falsifying purchase orders. Rob first called the district manager and asked if in fact the allegation was true; to his surprise, she readily admitted it was.

At that point Rob went to see his new boss, Phil Rouse, a big bear of a man, open and warm, with a great sense of humor. Phil was a good listener and absorbed Rob's information without interrupting. At the end of the story he jumped up from his chair, said, "Follow me," and led the way to the office of the general manager, Gene Mechlenberg.

They burst into Gene's office and Phil began telling the story almost before he and Rob could sit down. Gene heard him out, spun around in his swivel chair and took a plaque off his wall; across the top, the framed document read *The Principles that Guide Avon.*

Lebow knew about these principles but had assumed they were just another corporate decoration, as in so many companies; he never expected to see them consulted in deciding how to handle a business issue.

After a short discussion of the principles, all three men knew what needed to be done. Because she had admitted her actions, the erring district manager was suspended without pay for one month, and the press was advised of the incident. (Had she tried to stonewall her behavior, she would have been immediately fired.)

Do clearly stated values provide useful guidance for decision making? They did for Avon then—using principles that had been set down by the company's founder in 1903.

Virtually every business book beats the drum for the importance of a powerful statement of corporate values; why bring the subject up yet again? To make corporate values come alive, we need to go beyond just writing them down; we need to find the secrets of shaping a values statement that can guide *behavior*, and that can be applied in active, everyday use—key elements mostly overlooked.

The singular purpose of an organizational set of values is to help us understand who we are, why we do what we do, and what we aspire to. But we need to appreciate that intent, recognizing values are offered as guideposts for making the right decisions, not as a set of rules that will always give us the exact answer. And in fact, the Avon principles didn't give the three men a specific answer to the situation they faced. Those principles read:

THE PRINCIPLES THAT GUIDE AVON

TO GIVE TO OTHERS AN OPPORTUNITY
TO EARN, *in support of their happiness, betterment,
and welfare.*

TO SERVE FAMILIES EVERYWHERE
*with quality products that will bring satisfaction
to all users.*

TO RENDER A SERVICE *to customers that is highly outstanding in its
helpfulness and courtesy.*

TO SHARE WITH OTHERS *the blessings of deserved growth and business success.*

TO DEPEND WITH FULL CONFIDENCE
on the men and women of Avon, and on Managers and Representatives, for their individual and collective contributions towards success.

TO CHERISH AND RETAIN the friendly spirit of Avon.

In nineteenth-century England and North America, many more people than today felt a strong linkage between their work and their religious convictions. They sensed in a much more vivid way than we do today that through their work they were carrying out God's will (though they frequently interpreted this to mean that the more you prospered, the more you must be walking in God's favor).

Most often a company's values represent what the founder stood for and believed in—even when that runs in a thread back 50 or 100 years or more, as with the Avon example.

But even though values change slowly, in many cases they do change. Late in their careers, the founders of prospering companies whose efforts had made them rich frequently tended to become introspective, spiritual, and eventually philanthropic beyond the bounds of everyday charity. Andrew Carnegie, the steel magnate who was counted among the robber barons, later carved himself a place in the pantheon of leading benefactors of U.S. society through the conviction he eventually came to hold that it's a duty of the rich to distribute their excess wealth; acting on this belief, he set up a vast array of organizations for education and charity, and is best remembered for the great number of public libraries he funded and provided books for, all across the country.

John D. Rockefeller, Jr., son of the oil millionaire, served the interests of all nations by deeding the land in Manhattan on which the United Nations building now stands. Milton Hershey, the chocolate king, founded an orphanage and school that through the years has benefited thousands of youngsters. The vast art collection of financier and industrialist Andrew Mellon formed the basis for the National Gallery of Art in Washington, D.C. (Though Mellon modestly insisted the museum not bear his name, grateful Washingtonians to this day speak of the National Gallery as "The Mellon.")

Many of the early entrepreneurs became more values oriented as they thrived and increasingly got in touch with their spiritual side. Men and women like these frequently come only in later life to the shifting of values that turns them into great benefactors. But just as with the People Values addressed in the preceding chapter, the Business Values of an organization can also be remolded and reshaped by a conscious effort to create an environment that better serves the needs of the workforce and management.

WHY VALUES STATEMENTS HAVEN'T BEEN WORKING

In the 1970s, '80s, and '90s, it became standard for people starting a new organization to define their vision, mission, and goals. This became one of the first things the founders did, almost at the same time as ordering their stationery. And for established companies as well, sculpting a vision or mission statement became a hot item, lining the pockets of many vision-toting consultants and academics.

Yet it's turned out that there was an unrecognized flaw in all this, a disconnect: companies embraced their new values statements, trumpeted them proudly to employees, stockholders, and trading partners, and then quietly forgot about them except on corporate brochures. The statements became part of the new-employee indoctrination, they hung, framed behind glass, in the lobby, corridors, and an occasional office, but they brought little change to the company in the way people behaved.

THE FALLACY OF "CORE VALUES"

Values in business have been seen by traditional thinkers as one single "core set." That comes about in part because of the way we've been thinking about values and how the subject gets taught in business-school classes. The current thinking puts values at the center, the core, of an organization's philosophy. At first glance this appears to be a sensible description. But in the practical, workaday world, it's equivalent to burying the ideas under layers of other matter. Buried at the "core," the values have little impact, little influence over our day-to-day business behavior. This view is suggested graphically in what we call the *concentric model,* illustrated below.

Yet most people know from their own experience that they only have to walk through the door of a company, plant, or op-

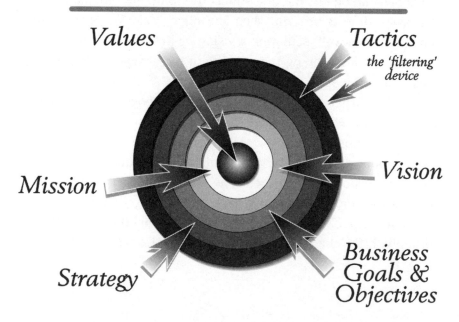

The Concentric Model of a Tactically Driven Organization

Values

Tactics
the 'filtering' device

Mission

Vision

Strategy

Business Goals & Objectives

eration, or wander the factory floor or the office corridors, to absorb the essence of the place. Plaques on the wall never tell us much; it's how people and the organization behave that tells the tale. We represent this view graphically with what we call the *molecular model*, which retains the same six elements, but with a dramatic difference: the values become the filtering device, and tactics are used to support values-based decisions, actions, and behaviors. (See the illustration below).

In the concentric model, tactics are the driving force; in the molecular model, tactics merely support the Shared Values of the organization, which include the dynamics of balancing Business and People Values.

Incidentally, these two graphics are closely related to the familiar "7 Ss of Management," first published in 1981 by Pro-

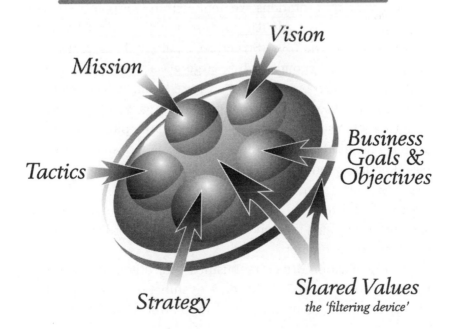

The Molecular Model of a Values-Driven Organization

Vision

Mission

Business Goals & Objectives

Tactics

Strategy

Shared Values
the 'filtering device'

fessor Richard Pascale of the Stanford University School of Business, and Anthony Athos, professor of Business Administration at Harvard University, in their landmark work, *The Art of Japanese Management*. The book contrasts a tactical organization—International Telephone and Telegraph, led by the iron hand of Harold Geneen—against a values-based operation—Matsushita Electric Co. The authors draw valuable lessons about short-term and long-term survival and success: at Matsushita, values were serious business. New employees were schooled on the Matsushita values, and the subject was systematically reintroduced throughout one's career—for example, staff members were required every two months to give a ten-minute discussion to their work group on the organization's values and how these values bettered society. A radical departure from International Telephone and Telegraph's dog-eat-dog work environment.

Pascale and Athos presented Shared Values as one element of the 7 Ss. We see this quite differently: Shared Values is the one central element that holds the other six together. In speeches and presentations, we demonstrate our point: we take a group of seven rectangular kitchen sponges, each labeled to represent one of the 7 Ss—Structure, Strategy, Skills, and so on. We put the seven sponges into a large glass water pitcher, and then add water, representing the Shared Values. The sponges expand as they soak up water, until they're packed together.

Just as the water causes the sponges in the pitcher to press against one another as they swell with the liquid, so in a values-based organization the Shared Values act, not merely as one of the 7 Ss, but as an element that connects all the other elements. And when we pour out the water and squeeze the sponges, they again take on the characteristics of disconnected elements in the pitcher.

An organization driven by business objectives and the bottom line alone can readily become dysfunctional because people have to "make their numbers," no matter what. The system then encourages people to stuff the channel, falsify orders, ship imperfect products. An organization's values work when its Business Values and People Values are in balance.

CRAFTING YOUR BUSINESS VALUES: THE ATTRIBUTES

So how do you go about crafting your own Business Values statement? Each value must have five attributes:

It must be *linked to the success* of everything you do.

It must *affect the success* of the organization.

It must be *controllable.*

> That is, it must call for something that can be assigned to one individual or a group—of employees, managers, or executives—who will oversee the efforts of the organization toward putting the value into action.

It must be *measurable.*

> How do you measure the success of achieving Business Values? Some suitable measures might be how you are seen in the marketplace as compared to your competitors; counts of damaged goods and on-time deliveries; complaints, repeat and referral business; customer longevity; the increase of sales to existing clients. . . .

It must be *aspirational.*

> A well-selected Business Value arouses people, calling for a high degree of effort, proficiency, or character. If it does not inspire, prod, or call to action, up the ante in the language you use. Elevate the call to action.

CRAFTING YOUR BUSINESS VALUES: THE PROCESS

The Business Values statement provides guidelines for how you will conduct business every day to be successful. The best statements are not product- or industry-specific: the best of them don't tell you what business the company is in.

You will be looking for an active, aspirational statement that is very individualized and company-focused. It needs to be strong and powerful enough to encompass all aspects of business operations, yet at the same time simple enough to be readily grasped by everyone.

As with so many other things in business, this effort isn't likely to go very far unless top management gets behind it, taking the lead in crafting Business Values.

One tip: we've found that the process works best when you bring in a third-party advisor to help you through the steps of crafting the Business Values statement—someone who is unbiased by any long-term involvement in the business, such as someone from one of your major vendors, or someone who has an affiliation with one of your senior executives, perhaps an associate from a trade organization. This advisor needs to guide you through the process, show you the way, facilitate . . . but leave company people to do the actual crafting and come up with the language. The result must be a company-derived statement: top management creates the initial stake in the ground, but everyone is invited to contribute to shaping the final form.

The actual process we recommend grows out of efforts at General Electric, back in the 1930s, when it was exploring ways to build consensus within its teams. We've modified its approach, adapting it to the needs of the Shared Values philosophy. While you may choose to carry out the process in some different way, our experience shows that the following approach can be highly effective.

Senior Manager Session

As pointed out, the process begins with the leadership.

Preferably at an off-site retreat, the senior management team drafts and refines an initial version of the Business

Values statement. It sometimes works best if the CEO comes with one or more ideas already drafted, as a starting point; however, if the senior managers are reluctant to challenge the CEO, or if the CEO operates in a strong command-and-control mode that keeps others from speaking out, starting with a statement already drafted by the CEO might defeat the purpose.

Through experience, we've developed one method that seems to work very effectively for getting active participation in these Business Values input sessions with the senior managers: each participant takes a stack of ten to 15 file cards (we use 4 x 6 cards, and give each person a single color), and writes on each card (preferably in large letters that can be read from a distance) a simple one- to five-word phrase that embodies one small concept. We call this phrase a "cept" because it is intended to be a single element of a *concept* that the person believes should be incorporated into the full Business Value statement; for example, "Hire people who are talented and dedicated" would be the text for two cards: the cept "Talented people" would go on one, and the cept "Dedicated people" on another. When everyone is finished (we allow about 15 minutes), each person selects what he or she considers the seven best ideas, and prioritizes these seven cards. Next, each person joins with a partner; from their 14 cards, each team agrees on the best seven, and again prioritizes. With large groups, the process continues in this way—forming groups of four, then eight, and so on—until there are about seven to ten groups remaining (that is, no more than about 70 cards still active).

Each group is now instructed to post its top card. (So that the cards can be easily moved around, we use a blank wall, and stick a loop of masking tape on the back of each card.) Then each posts its number-two card . . . and so on until all cards are posted.

Now it begins to get interesting: the entire group must create categories—for example, People, Quality, Service, and so on—and decide how to move the cards so each card is in the most suitable category. Encourage debate; this part can be especially enjoyable for everyone.

Once the cards have been categorized, each team takes one of the sorts or groupings of cept cards, finds a quiet place to work, and tries to fashion an umbrella statement that best represents the ideas on its group of cards. Each team then works to enhance its umbrella statement with a commentary line that elevates the statement into an inspiring idea; some people will write directive statements, others will write more philosophical ones; for example:

Umbrella statement: "Offer breakthrough products."
Commentary: "We will not seek 'footprints in the sand'; we will seek fresh sand to make new ones. We will then deliver the finest footprints possible with every step."

Through both formal and informal discussion coupled with "teachable moments," everyone in the organization will be able to understand and embrace the aspirational and visual nature of the "full concept" represented by these umbrella statements and commentary.

These are the steps we use to begin with a series of "cepts" and turn them into full concepts—in the same way that "Just Do It" doesn't take on its true meaning until it is framed in the context of Nike's philosophy.

When the group has arrived at a version of the Business Values it is comfortable with as a strong, on-target statement of the values that describe how the company functions or how the senior managers believe it *should* function, the statement will, in the steps described below, move on for consideration by others in the company. Each layer within an operation must have veto power. Without this level of trust, the rest of the organization will feel left out and dictated to, a return to the old command-and-control patterns.

Middle-Manager and Supervisor Sessions

We refer to the wording of the statement at this stage as a "work in progress"; it must not be looked on as a finished piece of work being presented for others to smile on and applaud (or more likely, *pretend* to applaud), but rather as a jumping-off point for discussion.

Mid-managers and supervisors receive a presentation on the newly defined Business Values (and, usually, the new People Values at the same time) at an informal workshop; when possible, the presenters should be some of the senior managers who played a role in shaping the new values. Those attending the workshop are then asked to examine and *test* the newly developed Business Values, make needed changes, and commit to the goal of creating a values-based work environment.

The group tests each values statement against the set of five attributes, as presented earlier:

Each Business Value must

- *Be linked to the success* of everything you do
- *Affect the success* of the organization
- *Be controllable*
- *Be measurable*
- *Be aspirational*

The best way to test the aspirational aspects of a Business Value is to ask the four key questions, also presented earlier:

Is each Business Value

- Good for the individual or customer?
- Good for the group, team, department, region?

- Good for the company or organization?
- Good for the community and environment?

Expect to encounter some contention over whether a particular suggestion truly represents a Business Value. Refer to the above criteria when a question pops up. A frivolous example may illustrate the point: your company is expanding into the ski-lodge business, and one executive has suggested that "Select locations on snowy mountains" is a value necessary to success. Test against the attributes: snowfall can be measured, but it can't be controlled (at least, not beyond the effects of a few snow-making machines, and no one has yet designed the machine that can create a 120-inch base). Conclusion: "snowfall" is an important business strategy, but not a Business Value.

Is "on-time delivery" an appropriate Business Value for a manufacturer? Yes. Match it up with the criteria above: it fits on every item.

Employee Meetings

Before Business Values can take hold in an organization, the rank-and-file members must arrive at the place where they believe in them, are ready to give them their support, and are ready to apply them every day; without that, nothing much will happen.

We achieve the understanding and commitment of employees by involving them in the process, through an opportunity to study how each value relates to everyday work and how it relates to each of the other values.

The procedure we recommend: Hold a series of companywide discussion meetings, preferably with no more than 12 to 18 employees at a time. In a small company, confine the meetings to the members of a particular department or work group.

A large organization, one with tens of thousands of employees around the world, might take as much as a year to conduct these meetings globally, and to feed the information back.

Examples of a Business Values Statement

When tackling a new task, everyone benefits by having some examples to follow. What are some existing statements of Business Values that we consider successful? The following are some of our favorites:

- U.S. Marine Corps: Everyone who serves in the Marines comes to take pride in the words *semper fidelis,* always faithful, a statement which covers everything from the cleanliness of your shoes to attempting the impossible on the battlefield.
- The City of Richfield, Minnesota: "One city committed to working together."
- Crysteel Manufacturing: "Take responsibility for being honest with yourself and each other."
- Augat Corporation: "Create an environment that develops people, provides an opportunity for self-fulfillment, and recognizes individual contributions."
- Gabbert's Furniture:
  "WOW the customer!
     Be fair
     Meet every commitment, keep every promise
     Partner in each other's development
     Simplify everything."
- Banta Digital Group: "Aggressively explore, integrate and utilize technology to achieve market superiority."
- Roberts Trane:"Do what's right—always!"
- President Homes: "The Golden Rule—Treat others in a manner that will create trust and respect."
- Bob Benson Saturn, Santa Rosa, California: "Listen! Take action! Then exceed customers' expectations."

- Excell Data Corporation: "We solve problems"; this is accompanied by a list of attributes to strive for, including flexibility, teamwork, good judgment, innovation, and learning from mistakes.

Keeping Values Fresh

A good customer calls and says, "I need that new desk delivered tomorrow." If you have a Business Value of satisfying the customer, each of your employees should know how to handle the situation; if they don't, then your Business Values, as well as your People Values, are no better than empty containers. If your employee tells the customer, "Monday is the day we deliver in your area," then your Business Values are not driving business behavior, blind policies are.

Many Shared Values companies print their values on wallet-size cards, listing the People Values on one side, the Business Values on the other. Permanently sealed in clear plastic, the cards are handed out to all employees. It's surprising how many people in a company are glad to get these cards, carry them in wallet or purse, and whip them out whenever there's a difficult issue to be dealt with. "Okay, we have a Business Value of Integrity that says we'll take responsibility for our actions. So I guess what I need to do is . . ."

Keeping Values Dynamic

Unlike People Values, which are much more universal and virtually eternal, Business Values need to be dynamic. Not that they change every few years: they appear to change quickly only when compared to the nearly unchanging People Values. It may be 15 or 20 years between changes in Business Values; still, every organization should examine its Business Values once every five years or so.

IBM provides an often-cited example. For years the company held a business value that had it providing both the questions and the answers: it itself defined what kinds of computers its customers would get, and the Management Information Services directors who signed the purchase orders waited for IBM to define the solution. When the desktop computer appeared on the scene, IBM got into the market but failed to realize that the whole dynamic was changing. In the past it had controlled the context of the buying decision. By the mid-1980s, the customers were no longer MIS directors, but the individual employees sitting at their desks, looking for real-time solutions. IBM should have been changing to a Business Value that offered real-time information at the desktop, just as PC software designers envisioned. Instead, the company held onto its old paradigm and its outdated Business Value. By the time it began to recognize the need for a shift, it had already lost control of the marketplace, and with it, its undisputed leadership. The happy end to the story is IBM's eventual awakening, revisiting and revising its Business Values, and showing every sign of returning to the enormously successful megacorporation it had formerly been. But to stay on top in the future means that it will have to continually revisit its new paradigms, altering its Business Values whenever it no longer fits the market. Never again will IBM be able to dictate to its customers.

The moral should be obvious: even though the process of reevaluating Business Values is time-consuming, it needs to be done once every few years, just to be certain you're playing the right game.

WRAP UP

A large part of the benefit in defining Business Values comes from involving your people in the process of evolving your own statement, tailor-made to your organization. (Which is, incidentally, a leading reason why it doesn't pay just to take another company's Business Values, adapt them a little, and call them your own. No one is likely to be worried about plagiarism here; "imitation is the sincerest form of flattery" is a more likely reaction. But you would lose the benefit of involving your own organization in the effort.)

Business Values define what your organization needs to do every day to be successful. At the same time, they are, in a sense, on the *outside* of the organization—they describe the ethics of how we behave, and form the standards that others will judge us on. And they also provide us the guidance for solving problems.

Well-functioning organizations are able to keep Business Values and People Values in balance—a necessity for any organization that hopes to remain competitive and viable.

Consensus Building

7

Leaders Have Lost
Their Followers

Lead, follow, or get out of the way.

A popular expression in the
U.S. Marine Corps

WHERE ARE THE FOLLOWERS?

Imagine this scene: you have a friend named Kirk, who is creative director at a major Midwest ad agency, and he's account executive—the boss—on the national Manticore Foods account. He joins you for lunch one day and he tells you the following story. (This is based on a real situation; the industry, company, and names have been changed.)

> "You think handling commission guys is tough? Try managing *creative* types—artists, writers, and other woolly-eyed geniuses. We were always throwing out new ideas and asking them to fast-track down some new alley where they've never been before. But Manticore is like another world. I knew what

they'd like and how to sell them when they weren't sure. If a creative meeting with the client wasn't going well, I'd step in and find a way to tweak the ideas, and leave them smiling.

"One day we're sitting at their big mahogany conference table, the agency people lined up along one side and the client people along the other. And they're not looking like happy campers. So I take out my pen and start rewriting the copy just like I've done so many times before. But I see Rudy, my chief writer, starting to squirm, so I try to put him at ease. 'You'll have to excuse me,' I say to cool him down, 'I guess I'm just a frustrated copywriter.'

"And he says, 'No—I'm the frustrated copywriter . . . you're the jerk.' He gets up and walks out—leaving me red-faced as Hell in front of the client."

Was this an issue of somebody unwilling to let the boss be in charge? Or was it something else?

The problem had nothing to do with whether Rudy was a good team player. It had nothing to do with whether Kirk was a good manager. The real question, which sounds like a contradiction, is, how could Kirk, the boss, have been a better *follower?*

Some people will tell you, "Everybody wants to lead. Nobody wants to follow anymore." Maybe it's not just their problem; maybe it's a problem with the boss, too. As a manager or executive, you should be asking yourself, "Do I ever listen to *other people's* ideas?"

One manager we know complains, "It's easy for the CEO; he's boss of the whole shooting gallery. He says 'Jump,' everybody says, 'Tell us when we can come down.' But out here where the work gets done—I say 'Jump,' and everybody says, 'We're too busy running to jump just now.' You can't just say, 'Follow me'; nothing happens."

If nobody wants to follow, maybe the reason has something to do with the fact that employees in their intended role as followers feel disenfranchised, and there seems to be no reward or payoff—because nobody trusts anybody else.

To become a good leader, you must become a good follower; the new follower is an *organizational citizen*.

This calls for a new approach to building consensus and building support around ideas. Central to these goals is what we call "the willingness to experiment"; the key is a willingness to try a particular approach *for some specified period of time*. The key question that's asked is, "Are you willing to experiment for a certain period of time?"—meaning the willingness to accept a decision even if you don't agree it's the best decision, give it your support, give it a full chance to prove itself successful, and only then, if it is clearly not successful, ask to revisit the decision and try something else.

"IF IT'S NOT MY IDEA, IT'S NO GOOD"

In the past, we welcomed strong business leaders like Henry Ford, Cornelius Vanderbilt, and John D. Rockefeller—entrepreneurial, capitalistic chief executives driving for market dominance, replaced after World War II by the command-and-control executives who were less entrepreneurs than they were industrialists and scientific managers in the postwar Peter Drucker mode.

Since the mid-to-late 1980s, the trend has been toward executives who operate in a much quieter way, with less bravado and more long-range thinking. Although few would like the label, this is in fact more in the Japanese mode of behind-the-scenes influencer and consensus builder.

At the same time, employees and managers have changed as well. Again the label will be rejected by many, but an accurate characterization for a large proportion is more adversarial—people who want their ideas to win at all costs.

The loss of paternalism, the demise of loyalty, have fostered individuals who model differently, and the model is patterned on self-centered individualism. As a result, many people, once they have tasted some decision-making and self-expression, lose the capacity to follow. Those who make up the population of Generation X are not good followers, and perhaps it's not all their fault. Just as being a man in today's society requires a lot more awareness and sensitivity, so being a follower today—of either gender—is a lot trickier than it used to be.

It's unlikely we'll ever go back to command-and-control. So the person who would be a strong leader today faces the challenge of becoming a *facilitator* who can gather a new style of follower: people who are the organizational citizens of the future. The citizens in the new Shared Values organization fully understand the eight key elements to their participation and responsibility in a new partnership with their leaders.

GATHERING THE NEW ORGANIZATIONAL CITIZENS

The new facilitating leaders are those who learn how to enable their followers to become new citizens, and the success of a new citizen can be measured against eight elements:

The Eight Key Elements of Gathering Followers to Become New Citizens

When people become the new citizens, they:
- Become contributing members of their group
- Openly seek and encourage others' opinions
- Accept that different viewpoints are ultimately helpful
- Are comfortable stating an issue in their own words
- Express their views about an issue
- Are willing to experiment for a period of time with someone else's idea even if they don't agree

- Understand that all group members must support the final decision
- Are willing to take responsibility for communicating and implementing the final decision

Note that these goals describe a citizen in terms of becoming a willing part of the group. In particular, being willing to experiment with someone else's idea even when we think another idea is better means rejecting the common and familiar attitude that says, "If you win, I lose; if I win, you lose." Instead, everyone in the organization needs to become committed to a new attitude that says, "If it's a win, we *all* win."

Taken together, these important goals describe the process of consensus building. How do you go about building consensus in the organization?

BUILDING CONSENSUS

How do you put into action these eight keys for gathering followers and building consensus? Some suggestions:

1. Be a contributing member.

Recognize that in creating and being open to new ideas, you must take responsibility to step forward and express your ideas. Bystanders have no place in a Shared Values environment.

This puts a responsibility on the shoulders not just of the group leader, but of each group member. The goal: a group of citizens who are contributing members who actively speak up.

2. Openly seek and encourage others' opinions.

Good decisions grow out of a process that begins with gathering information as a first step, and then seeking other people's opinions. Full support only comes when people understand what the decision was based on and have had time to be part of the discovery process.

Once a direction for a decision is proposed, all those who will be affected must be brought on board. You must spend the necessary time at this stage in the building of consensus, or you will be forced to do it later on, to overcome the inevitable pushback.

3. Recognize that different viewpoints are ultimately helpful.

Rather than detracting from the progress toward reaching a decision, different viewpoints can add spice and life to the discussions. Without these contributions, choices become limited and opportunities diminished.

One caveat: some group members will defocus and divert the group—through lack of skill in participating, or as a result of the "If you win, I lose" attitude. The challenge to the group is to welcome diverse opinions, yet block those ideas that get in the way of moving ahead. This is often a tough call requiring good judgment—to keep moving in a positive direction, yet not stifle valid views just because they're different from the favored opinion.

Ask yourself these questions: Do I accept different opinions? Do I help create an open atmosphere? Would others say I'm open to new ideas? Have I thought of ways to improve openness in meetings? Have I overcome rigid thinking?

4. Learn to be comfortable stating the issue in your own words.

In a group that's working together effectively, everybody participates. That doesn't mean everyone has to have the floor for five minutes at every meeting; it doesn't mean everyone has to offer an opinion on every issue that's raised. But it does mean that, over time, everyone in the group should be making regular contributions.

Though you don't need to have an opinion on every

subject, when you do have one, you have a responsibility to speak out. If you stay silent, you give up your franchise— diminishing both yourself and the organization.

A lot of people, feeling they may be unpopular, prefer to keep quiet, claiming they don't have enough information to voice an opinion. The group needs to help these people understand they have an obligation to gather or become familiar with enough information to grow into a contributing member of the group.

On the other hand, sometimes people are reluctant to speak up for a very valid reason: the environment may not be safe for voicing an opinion. When this happens, the organization needs to be open enough that people can speak about the problem and expect that changes will be made.

5. Make sure you express your opinions about the issue.

We have an obligation to express opinions, and to do so appropriately.

6. Be willing to experiment for a certain period of time with a solution, even if you don't agree.

A lesson from the best organizations: people may have their disagreements, but once an issue is decided, it's wholeheartedly supported by the entire group or organization. In the early 1970s, the Nordstrom chain looked at expanding from a highly regarded shoe store into women's clothing. (It's not so much the Nordstrom management success as its values that explains why it is so often used as an example in business books.) The decision to expand followed this very principle: one of the three brothers had a vision that included moving into women's clothing; the other two adamantly opposed the idea, arguing, "How can we expand into an industry that we have no real background in?" Yet, the one brother was able to ask the other two to experiment for a period of time.

The brothers joined hands in the effort, working side by

side to make it work. In the beginning it was rough going, and it wasn't until the fifth year that they were enjoying great success. They had seen it through because two of the brothers had been willing to experiment for a period of time with something they did not initially support.

The real key lies not in making a decision—that's the easy part; the difficult part is getting the support and commitment to find out if it will work.

7. Understand that all group members must support the final decision.

Even when you've been careful to build consensus, you need to be prepared for some pushback after the fact. Don't reject the pushback and objections; listen to them. However, you or your team is under no obligation to change the decision.

After you've listened, acknowledge the information. Then you have several options: You can respond immediately to the new ideas, modifications, or requests. Or you can reserve reacting to their comments, telling those who oppose your decision that you need to think about their remarks. Or you can ask them to support the existing decision with the provision that it is an experiment and if it doesn't work, you'll be open to other approaches.

8. Be willing to take responsibility for communicating and implementing the decision.

Once an organization is successful in moving away from the old mode where workers and supervisors merely carried out orders and ideas generated at the top of the pyramid, where employees were passive, compliant bystanders, the roles of communicating and implementing become essential.

Today, each member of the staff, regardless of their role, needs to understand why things are done the way they are—not blindly accepting the status quo but being a partner in creating an evolving approach to business operations. In fact, if enough members of the organization don't want to get involved, they become a liability to the entire operation.

Organizations can no longer tolerate the bystanders who remain on the sidelines, keeping silent but creating tension and mistrust.

WRAP UP

The essence of gaining values-based consensus lies in helping leaders and members of the organization learn how to become the new values-based citizens who find ways to make ideas work rather than workers just concerned with whose idea is ultimately accepted. This requires a new philosophy centered on participation, communication, support, and experimentation, combined with the process to pull it all together.

An organization that has an agreed-upon consensus philosophy and set of behavioral standards can do things an ordinary organization can't. Individuals will feel a partnership with the accepted decisions. They will reserve the right to disagree, but will step forward in the spirit of a "can-do atmosphere" because they know that sometimes the group will be working to make one of their own ideas *fly!*

8

Privileges and Rights

There would be fewer arguments if more of us tried to determine what's right instead of who's right.

Anonymous

THE NATURE OF RIGHTS

Virtually all advanced societies today guarantee their citizens an extensive set of rights—including many once reserved to the "high born," the landowners, or, as in ancient Greece, the freemen.

The United States in particular is a nation founded on the principle of the rights of the individual, a heritage in which we take a deep pride. The Bill of Rights forms a contract between the nation and the people, one that many other nations have admired and emulated. (Even the Magna Carta, looked on as among the most exemplary guarantees of rights, sought to protect British barons from the power of the King, but gave no thought to the rights of the ordinary citizen.)

Yet, as management consultant Keshavan Nair, Ph.D., has pointed out (in "A Clue from Gandhi," *Sky* magazine, May 1995; Dr. Nair is also the author of *A Higher Standard of Leader-*

ship: Lessons from the Life of Gandhi), basing conduct and policy on a concept of rights eventually creates a society "driven by advocacy, leading to a loss of community and reducing the motivation to work for the common good." In other words, we begin to use rights not merely for preserving our own quality of life, but as a weapon against other people and against the institutions of the society.

So we've become a society where many of us will at times insist on what we see as our rights even if it means trampling on the rights of someone else. We argue from an advocacy position—as employees, as customers, as shareholders. Even criminals remind us of their rights. We've become a litigious society in which handshakes have been replaced by lawyers filing lawsuits, significantly adding to the costs of all our goods and services.

Competition, in its place, is good, and healthy, and fruitful—one of the elements that has made the American society such a triumph of enterprise and prosperity. But we've reached a time when competition sometimes turns into a dark force of destructive power, damaging us as individuals and damaging the fabric of our society.

A RETURN TO RESPONSIBILITY

In the essay cited above, Dr. Nair goes on to present the views of Mahatma Gandhi, the great leader who won independence for India from the British through a policy of nonviolence. Gandhi believed that "a commitment to personal responsibility, not insistence on rights, should govern conduct and public policy," and that if you start by emphasizing the *duties* of man "the rights will follow as spring follows winter."

Insisting on our rights, we've lost the sense of responsibility in much of our daily life. But when we look for them, we find that examples still abound. Every time you slow down to let another driver move into your lane of traffic, even when he didn't bother to use a turn-signal, you're being guided by your responsibilities instead of your rights. Every time you handle a problem even though it resulted from a co-worker's mistake . . . or help a child with homework though it means missing your favorite TV show . . . or do volunteer work or give to charity—you're acting on an impulse of responsibility instead of an impulse of your individual rights.

Farmers in many communities still act like responsible neighbors—pitching in to help one another in planting and harvest seasons, even though at market time they'll be battling the same neighbors to get the best selling price.

Where is that collective spirit in the workplace? Not entirely vanished, to be sure, but not half the force it once was. A Shared Values approach provides a road back to a sense of individual responsibility. And this must be a two-way street: it calls for a renewed sense of responsibility not just on the part of the employees, but on the part of the organization as well.

A NEW FOCUS ON RESPONSIBILITY IN THE WORKPLACE

The basis of the relationship between the employee and the organization used to be the promise of job security. In today's world, where most companies can no longer extend loyalty in the form of job security (and so can no longer expect loyalty in the form of a long-term commitment from employees not to take their skills to some competitor down the street), a new set of responsibilities needs to be found on both sides of the equation, one that will replace the set lost in a world of global competition, short product cycles, and shrinking layers of middle management.

In a Shared Values environment, leaders begin to recognize their responsibilities—for example, by putting their people first; by articulating the relationship between the organization and employees (a manager's greatest job is to create the conditions for success); by providing truthful information about the company's future and the future advancement opportunities for employees.

To pick just one prominent example, responsibility to employees, as well as to investors and stockholders, also means committing to long-term investment so the company will have a decent chance of still being around when today's newest employee is ready to retire. Company managements have become so wary of long-term expenditures that it's headline news when an Intel announces plans to construct a billion-dollar semiconductor fab (industryspeak for a computer-chip manufacturing plant), or a GE announces its billion-dollar project to develop that jet engine which won't begin producing income for 25 years. (And the willingness to take this kind of risk to insure future stability is certainly a leading reason that GE is one of the few major companies to survive intact from the turn of the twentieth century.)

And what about the responsibilities on the part of employees? At the top of the list is an item that benefits the employee as well as the organization: the responsibility to continually grow in knowledge and ability—learning new skills; taking courses—within the company or outside; mastering jobs other than one's own; gaining in ability, command, and stature.

WRAP UP

At Johnsonville Sausage (a company cited at some length in upcoming chapters), no employee gets a raise on the basis of longevity; employees earn additions to their pay by mastering some new skill. And for frontline workers, it goes beyond just learning how to run another machine—it may mean learning bookkeeping skills or the art of developing a marketing program for a new product.

So Johnsonville employees don't expect an annual raise as their right; the only annual issue is "What have you done to improve your value?" There is a lesson here for other companies. But the same logic needs to apply to the organization, as well. The question each company needs to ask is: "What steps is this organization taking to improve the working conditions of our employees, and the support we give our employees, to contribute to their short-term and long-term growth and competency?"

And finally, we've all heard that the employee starting a career today will hold some five or six very different jobs during her or his working life. Continual learning that will make those job shifts less jolting is a responsibility of the employee to pursue, and the organization to encourage and support.

In the following chapters, we'll see specific techniques for making these things happen.

Values-Based Responsibility-Taking and Decision-Making

9

Letting Go: Giving Responsibility Back to the Rightful Owners

---

My job is secure. They can't fire a slave.

Sign on an office wall

Reward excellent failures. Punish mediocre success.

Phil Daniels, participant in a Tom Peters seminar in Australia

---

When Ralph Stayer sent a research team in to conduct a survey of his employees, he was startled by the results—which reported very little to make him smile.

Stayer was then head of a Wisconsin-based family business, Johnsonville Sausage. You don't expect the ah-ha's of a sausage maker to appear in the august pages of the *Harvard Business Review;* he not only had his experiences published there (November/December 1990) but offered up a refreshingly different view of the role of manager.

What shocked him about the survey results was the news that his typical employee saw nothing at Johnsonville but a place to earn a salary. Stayer wanted committed people who cared about what happened to the company, and found instead that he had a workforce made lup of clock-punchers.

After a lot of soul searching, Stayer decided the problem wasn't with the employees but with himself. "I had made all the decisions about purchasing, scheduling, quality, pricing, marketing, sales, hiring, and all the rest of it."

How did he solve the problem? Hint: Stayer titled his *HBR* article "How I Learned to Let My Workers Lead."

MANAGERS SHOULDN'T MANAGE

How do managers and corporate executives typically view their jobs? One gag definition describes a manager as someone who has hired so badly, he thinks he has to keep close watch on the poor devils. Employees only became uncommitted when the manager takes away responsibility. But remember Deming's remark about deadened employees: "Did you hire them that way, or did you kill them?" Employees *become* detached from their jobs when managers refuse to share responsibility.

Who is responsible for the performance and productivity of the people in a work group? Most managers will tell you it's themselves; just like the captain of a ship, the manager is responsible for the results of her group. And then they'll sit there and complain that their employees don't take responsibility. So what is it they want the employees to take responsibility for— the paper clips?

The bottom line is that managers won't relinquish responsibility because they're afraid they'll be left with nothing to do, yet will continue to be held accountable for results. So they react by dominating and controlling their people. In the typical company, employees lack a sense of being in charge of their own work because of the way their boss manages them . . . and the boss suffers just as acutely because of the way his boss manages him. The result: as organizations have become less responsible for their people, people have become less responsible for their work. The more impersonal work becomes, the less responsible we become.

MACHINES DON'T CARE ABOUT PEOPLE

At the same time, we're automating processes in a way that makes it harder and harder for people—customers as well as employees—to interact on a personal level. A Rupert Murdoch or a Donald Trump could get a bank loan based upon his name and reputation; for the rest of us, a set of numbers gets cranked through a computer—years on the job, years at the same address, renting or buying, total household income—and the machine gives a "Yes" or "No"; we may get the answer from a human but we know a machine has told her what to say. In manufacturing, so many of the lathes, stamping presses, and spot welders are automated that the operators spend much of their time as observers and monitors. In some automated factories, workers who used to grease the gears and adjust the tension now sit in front of flickering screens in a control room; they may gain some satisfaction from wearing a clean shirt instead of coveralls, but deep down they know that most of the time they are nursemaids to the fancy electronics that are really running the system.

A customer calls our offices and is greeted by an electronic voice offering a series of branching choices, and the customer may complete a transaction ("fax-back," for example) without ever speaking to a human; even pressing zero for the "attendant" often gets an electronic voice which tells the customer she has made an invalid choice.

On the airlines, there's less legroom, less food, the seats are narrower, the cushions thinner, and thanks to staffing cutbacks, the in-flight attendants have little time to be pleasant. Ballparks are being designed with larger capacities, sticking the fans farther away from the action. Teaching relies more and more on the handy videocassette and the clever interactive CD program; the live teacher becomes a standby for anyone having a problem—efficient, yes, but far less personal.

During the business day, in thousands of small ways, as we learn to trust machines instead of people, we lose the sense of human contact. This despite the universal need, the yearning, for personal interactions.

Viricon, an international company based in the Midwest, which is a leading fabricator of architectural glass products, did a study of its most successful employees under age 25. What it found was unexpected. The top employees appeared to share only two factors in common: most came from a two-parent family . . . and most had played a team sport—in the process learning how to deal with others and to be personally accountable.

We'd like to turn the workplace into something akin to the excitement, cooperation, and shared responsibility of a basketball or soccer team. What happens if we just tell your work group, "It's yours—you make the decisions," and walk away?

GIVING BACK RESPONSIBILITY

When Ralph Stayer finally came to accept that his employees lacked commitment to Johnsonville Sausage because they had no authority to make decisions or to control their own work, he took a radical step: he placed full authority in the hands of the management team, giving them responsibility for their own decision-making.

We argue here that you and I and virtually all employees want to be given responsibility, so you might reasonably expect this story to have a happy outcome—right?

Hardly. Stayer had gone, in his words, "from authoritarian control to authoritarian abdication." The managers struggled to meet the challenge, but they had been trained in a company boot camp where the number one lesson was "Take all matters to the CEO for a decision." Their experience at the company had taught them little about taking matters into their own hands. Stayer eventually realized his managers were not focusing on making the best possible decisions, but instead trying to guess which decision Stayer himself really wanted them to make. They were trying to become mind readers instead of decision makers.

Stayer eventually solved the problem (but not without great pain; see the next installment of this story, in the following chapter). It took two years before he realized that people who've been working in a command-and-control environment have to *learn* how to take responsibility.

How do we help employees take responsibility for decision-making, interacting, and measuring? Like Stayer, most managers have only the dimmest notion that the effort is even desirable, much less how to go about it.

Part of the answer lies in allowing people to become closer to the business, in much the same way we want people to become closer to the customer. We've already talked about opening the books, which is one of the ways of doing this. But the essential ingredient is what we call "giving responsibility back to the rightful owners."

THE "RIGHTFUL OWNERS"

The story of cops and robbers in New York City doesn't sound as if it would have a place in a management book, yet it's highly instructive. The history of law enforcement in New York parallels the experience in many other places: the cop walking a beat is not nearly so easily supervised, controlled, and accounted for as the cop in a squad car. The precinct captains always worried that their foot patrol officers could be spending much of the shift in a coffee shop, pool hall, saloon, or topless bar—with no one the wiser. Squad-car officers, on the other hand, are always within arm's reach of the radio; if they leave the car for a crime-scene investigation or a sandwich, they have to tell the dispatcher what they're doing.

Over the years, bureaucratic thinking won out. The New York Police Department, like many others, pulled cops off the street and plunked them into cars, and meanwhile built up layers of blue-uniformed management at headquarters. It turns out, though—as the NYPD found out when it finally reverted to the earlier practice—a cop on the beat is a tremendously effective deterrent to petty crime. He (or she—the percentage of females on the police forces of American cities is up dramatically over the past decade) is a visible presence, and knows by name or at least by sight many of the citizens and shopkeepers, and many of the hangers-on, loafers, and drug dealers. Even honoring the various protections our system grants all citizens, the cop still has authority in many circumstances to frisk people, and in high-crime districts of New York City, a lot of the bad guys have stopped carrying guns because of these unavoidable pat-downs. And the statistics are now clear: if you can cut down on petty crime, you eventually cut down on the serious crime as well—the number of homicides committed annually in the city plummeted an astounding 57 percent in just six years. It might be said that the new thinking reflects a goal of arresting crime instead of just arresting criminals.

Still, it's on the management side of the equation where the most dramatic changes have taken place. Precinct crime statistics used to be sent to headquarters a few times a year. Now they're sent weekly on computer disk and pored over from every perspective.

Twice a week, every precinct commander meets with the top brass in the headquarters "war room" for a grueling session called Compstat (computer statistics). With only a day or two advance notice, a selected commander and his or her key aides stand up to present the statistics and explain what they're doing to improve the subpar numbers. Journalist David Anderson, allowed to sit in on one of these sessions while writing an article for the *New York Times Magazine* (2/9/97), heard a precinct narcotics supervisor grilled over his numbers. "Your arrests are low on Saturday. Aren't they selling dope on Saturday?"

No longer a system run by headquarters' bureaucrats far removed from the streets, responsibility now rests heavily on the shoulders of what we refer to as the rightful owners—the people who are doing the actual work and should be held accountable for it. And they *are* being held accountable: a commander whose numbers and explanations don't measure up two or three times in a row will have his command taken away.

The system is working—New York crime statistics continue to drop, and other police departments are coming by to learn the lessons. But it hasn't been easy; changing a culture never is.

If you want employees in your organization to begin taking responsibility, you need to start seeing them as the rightful owners, the people to whom these responsibilities actually belong. You need, as well, to begin seeing them as people entitled to certain inalienable rights. These are rights they had when you hired them; you thought these were people worth hiring because you thought they'd be "responsible." But once they came to work, they butted up against policies, procedures, and management and found responsibility snatched away.

So how do you start to change the patterns? Try moving away from the table.

MOVING AWAY FROM THE TABLE

This phrase implies pushing your chair back and becoming an observer as your work group takes the reins and begins making decisions on its own. But Ralph Stayer's experience is a warning: you can't expect a fruitful outcome if you do it all at once in a sudden shifting of years'-long patterns. As Stayer learned, managers who move away from the table as a sudden change in behavior are not exhibiting good judgment; they are in effect abandoning the group. We'll see in the following chapters how to lead others to embrace responsibility-taking and accept the new challenges enthusiastically.

Underlying all this is the philosophy that groups make better decisions than a single leader. Stayer expanded on the notion in the book he co-authored with James Belasco, *Flight of the Buffalo*. The witty title suggests a compelling metaphor of two groups and two choices.

The buffalo herd is organized around a leader of enormous strength that plods forward while the rest of the herd, blindly and unchallenging, follows wherever it leads. If something happens to the leader, the herd stands around in disarray, unable to figure out what to do next. (A fact discovered by the early hunters, who would drop the leader and then pick off the rest of the herd at leisure, a too-easy practice that continued until the species was nearly extinct.)

Compare that to a flock of geese. Geese have a novel way of leading. All birds in the group share the mission and destination. As they fly along, each goose assumes responsibility and takes over leadership in a rotating fashion: whenever the lead goose begins to grow tired, it drops back in the formation and another takes over. Each goose assumes leadership and shares the power. The flock is not vulnerable like the buffalo because the fundamental operating system is different.

Waterfowl Magazine, published by Waterfowl U.S.A., extended the description of the flock-of-geese dynamics in its issue of December/January 1991–92, which we've expanded by interpreting the metaphor to the workplace.

Lessons from Geese

As each goose flaps its wings, it creates an uplift for the birds that follow. By flying in a "V" formation, the whole flock adds 71 percent greater flying range than if each bird flew alone.

People who share a common direction and sense of community can get where they're going quicker and easier because they're traveling on the thrust of one another. Shared Values is based on including all members of the organization and excluding no one.

When a goose falls out of formation, it suddenly feels the drag and resistance of flying alone. It quickly moves back into formation to take advantage of the lifting power of the bird immediately in front of it.

If we have as much sense as a goose, we stay in formation with those headed where we want to go. We're willing to accept their help and give our help to others. The whole basis of mentoring is the value and power of the "lift of the wings" from the person who mentors us.

The geese flying in formation honk to encourage those up front to keep up their speed.

We need to make sure "honking" is encouraging. Giving credit where it's due is the key to supporting, enabling, and empowering ourselves and others. Everyone's job in a Shared Values environment is to help contribute to creating the conditions for the success of all. In groups where there is encouragement, the work effort is much more productive.

When a goose gets sick or wounded, other geese drop out of formation and follow it down to help and protect it. If it becomes able to fly again, the group catches up with the flock or launches out with another formation.

If we have as much sense as geese, we'll stand by each other in difficult times as well as when we're strong. Without loyalty, strong organizations are weakened and made vulnerable to outside forces. Loyalty must be an important element in every Shared Values environment.

For yourself and your work group, do you operate like geese, or like buffalo? Do you want to share leadership, or face extinction?

WRAP UP

We need to accept that it's an important strategy as well as morally correct to ask people to be held responsible for their actions. But this requires that they be able to take the responsibility. They must be willing to become full citizens, fully participating in the organization. And, with your permission, they must be able to rewrite the rules.

How you'll go about doing all these things is the subject of the next three chapters.

10

The Eight Keys to Responsibility-Taking

Managers do things right; leaders do the right thing.

An often-quoted maxim of Peter Drucker

Several years ago we were called in by a firm in the business-forms industry, a leading company more than 100 years old, with annual revenues upwards of $1 billion. But its sales had been flat for three years.

To take the pulse of the company, we conducted a personal survey of 250 people with sales-related functions from shelf stockers, to vice presidents, to the Executive Council. In each case we explored their perceptions around three areas:

- How do you define the term "empowerment" as it relates to you and to your job?
- Define what comprises your sales team.
- What do you need in the way of tools, processes, policies, or support to help you increase your sales?

On the definition of empowerment, people didn't answer the question as asked but instead answered the flip side, in terms of disempowerment—what they were *not* able to do. At each level they were asking for what has historically been defined as their boss's job.

On the sales-team question, not one of the people at any level saw the person they reported to as being on their team.

When the "what do you need" results were tallied, here's what we found: the shelf stockers said, "Get the district manager off my back and I can do my job."

The district managers said, "Get the regional manager off my back."

The regional managers, area vice presidents, and corporate vice presidents each said essentially the same thing: "Get my boss off my back."

And the Executive Council? They said, "We don't understand—we've given these people all the resources they could reasonably need. Why can't they improve sales?"

We agree that the old definition that good employees were compliant and did what they were told now needs to be replaced, but by what? Our answer is that all employees must become citizens of their organization, or even leaders. If you have 300 people in an organization, you need to have, not 30 leaders, but 300 citizens.

And that's why you need to get rid of the rulebooks and procedures manuals. Each employee must take back responsibility, and make the rules that they themselves will follow. The Nordstrom chain gives all employees a small plastic card about the size of a small piece of paper. On one side it says "Employees' Manual." On the other side it says: "Rule #1: Use your good judgment in all situations." And then it says, "There will be no additional rules." That constitutes the entire employees' handbook, and in large measure it explains the extraordinary success of Nordstrom for years, and the success of other businesses that follow its philosophy.

You can't throw away your own company's policies and procedures manuals overnight, as we've seen—not until you've helped people learn how to take back the responsibility that the organization has taken away. Few people see themselves as leaders, but everyone can see himself or herself as a citizen.

A Chinese proverb says, "Only hire people you trust. Once you've hired them, trust them." Managers will be extremely well served by coming to believe that people want to do the right thing. This is the basic premise of the Shared Values Process Operating System—that people are going to do the right thing.

LEARNING TO TAKE RESPONSIBILITY: THE EIGHT KEYS

So how do you get people to embrace responsibility-taking in a way that sticks, even when the pressure is on? Companies all over the world, hundreds of them, are adopting the ideas of the Shared Values environment, and are finding that once they start from the premise that people want to be the best they can, then their employees go way beyond expectations.

Job number one for all managers and leaders is to create the conditions for superb performance. W. Edwards Deming suggested a standard for this, which he termed "extraordinary predictability"—you don't want performance that sometimes measures up, and sometimes doesn't. For us the term "superb performance" means how consistent we are with our values, our standards, our philosophy. And whether everybody, throughout the entire organization, plays by the same set of standards.

Over the years we've developed a Responsibility-Taking process that helps people grasp the key ideas and put them into practice. We introduce the topic to managers and executives by means of a demonstration using eight marbles. Imagine these eight marbles are on the table in front of you now; pick them all up in your right hand—representing that you, as manager or leader, have responsibility for all eight items. After reading the description of each of the following, if you have given responsibility to your group for that action, transfer one of the marbles to your left hand.

But beware: if you've given the group responsibility yet retain the right to overrule them, you've not given them full responsibility, and *the marble stays in the right hand.*

Read the description of each of the eight keys, review the guidelines, decide how you're currently handling the situation, and determine which hand the marble belongs in.

1. Planning

Do you allow your people to take an active role in gathering information, strategizing, planning, executing the plans, and measuring the results?

You are succeeding in this area if your people do their own planning, without pressure from you. In the story about the business-forms company at the beginning of this chapter, employees at each level believed they were continually being subjected to roadblocks imposed from above, in many cases imposed from far away by people at corporate who were out of touch with the real situation and needs in the field. Most of the employees in the survey felt close enough to the business to believe they could set local priorities that were realistic, immediate, and would increase results. Never was there a hint of disloyalty—all stated a firm belief in both the products and the company. But they felt great frustration at not being able to take risks, try new things, or reject policies that seemed clearly inappropriate to their own local situation.

In most organizations, much of the policy setting should be regional or local, not national; generally speaking, national policies fit no one well. Centralized planning fails because the people who plan are not the ones who execute. You can frequently predict if a plan is going to succeed by looking at who's doing the planning and who will be carrying it out. If they are two different groups, it's likely the plan will underperform.

Guidelines for Planning

Encourage people at all levels to take an active role in gathering information and strategizing future actions.

In your planning process, include the people who will be affected, both directly and indirectly *(if this represents a large group, use a representative sample).*

Give the people who establish a plan responsibility for the execution and measurement.

If people reporting to you are at more than one location, decentralize the planning to achieve full responsibility-taking.

Transfer one marble to your left hand if you are already following most of these guidelines.

2. Priority Setting

Do you allow people to set their own priorities?

To be able to do this, each individual needs to have access to "the big picture." People can only set priorities when they understand fundamental business skills. Here, education takes on a new meaning. It becomes the responsibility of the organization to educate and challenge everyone to be involved.

One of the things you see early on in the transformation to a Shared Values environment is that many people feel uncomfortable setting priorities. They find themselves faced with an unfamiliar situation, bringing concerns that they will be held accountable, that they don't really have the big picture. And this requires them to think at a high level—which in most organizations people are not often asked to do.

So don't expect that everyone will at first be eager to participate in setting priorities. Some have the attitude, "If you want
me to do this, pay me a manager's salary because that's a manager's job."

Guidelines for Priority Setting

Help people learn to become comfortable with
setting priorities.

Consistently share "the big picture."

Invest resources to bring your people up to
speed on fundamental business skills.

If you are already following these guidelines, transfer one of
the marbles to your left hand.

3. Removing Roadblocks

Have you given your people the training and the authority so
that they can remove their own roadblocks?

In most organizations, people don't remove roadblocks; they
circumvent them, ignore them, or, more likely, are stifled by
them. The roadblocks at the business-forms company in the
earlier story were largely handed down from corporate. Imagine the power that the organization could have gained if people
on the front lines had been able to remove those roadblocks by
dealing directly with those who set them up.

"Skunkworks" projects—the very name suggests an operation set up to keep unwanted bureaucrats and rule makers at a
distance—are something an organization resorts to when, because of an impossible schedule or a difficult engineering, production, or creative challenge, there's a recognized need to give
an operation special permission to circumvent the usual policies and procedures. Isn't something being missed here? If the
normal environment in the company is so stifling that a high-
profile operation has to be given special breathing room, what's

happening to the rest of the organization—all those day-to-day efforts that *aren't* being protected? With things so badly out of kilter, the only solution it can find lies in setting up a skunkworks . . . while everything else suffers under the weight of the burdensome procedures manuals.

Oppressive, restrictive rules aren't confined to old-line industries and old-fashioned companies. Chuck Yeager, one of the foremost aviators of all time, had "the right stuff" . . . but not what NASA officials defined as the right stuff required to become an astronaut. So he broke rules and broke the sound barrier, but never put on an astronaut's space suit, never flew a space mission, and never received the accolades of his nation.

Business leaders face difficult challenges when trying to distinguish between the rules that get in the way of better performance by the company's people, and those without which the organization would quickly descend into anarchy. Aetna Insurance allows its employees to order office supplies, computer accessories, and similar items up to a certain price level through an online purchasing system, *with no approval required;* (there are reviews of the purchases, of course, and those few who abuse the privilege are quickly identified and dealt with). In one wave of the wand, Aetna did away with an entire layer of bureaucracy devoted to reviewing requisitions for ballpoint pens and paper clips, and at the same time removed one persistent roadblock from its employees.

The organization has a responsibility to remove roadblocks, but so does the individual. Each person in the organization regularly faces three alternatives: ask permission to remove a roadblock; go around it and make excuses after the fact; or be disempowered. And when people need to ask permission, then it's the manager who is removing the roadblock.

For an organization to approach the level of effectiveness it's capable of, its people need to actively strive at removing roadblocks themselves as part of their everyday approach to work.

Many of a company's rules and procedures clearly can't be dropped—no firm would give every employee the unilateral authority to hire or promote, for example, or give every manager the right to contract for a new factory. In between the extremes of buying paper clips and building a new plant lies a vast gray area demanding flexibility and judgment.

These guidelines will help point the way:

Guidelines for Removing Roadblocks

Be on the lookout for red tape, policies and procedures that stifle, staff versus line issues, the "not invented here" syndrome, gatekeeping, and hidden agendas. Develop an active early warning system against all roadblocks.

Do your part in leading the culture of the organization to become one that encourages people to remove their own roadblocks.

Help your employees understand they must be able to remove roadblocks before they can take responsibility.

Push for your organization to redefine the manager's role, so that managers are encouraging staff and associates to remove their own roadblocks.

Put one marble into your left hand if you are already following most of these guidelines.

4. Encouraging Creativity

We hear "creativity" and most often the artistic sense of the word is what first pops into mind. Creativity in business of course encompasses a lot more than the output of the advertising copywriter and the Webmaster who designs the company's Internet site. Creativity also means self-expression—the ability of each person to make choices.

When creativity is missing from our work, we suffer a loss of inspiration, energy, enthusiasm, fun, loyalty, engagement, and commitment.

Yet, without meaning to, without even recognizing what's happening, most organizations over time manage to filter out all the creative elements from each job, leaving people to go through work in a lifeless, joyless, uninspired shuffle. A telltale sign: how many of your people know the number of days until they retire?

Not that every phase of a job has to be creative—but there must be some creative parts to every job. It's the creativity that engages us in the work. Even a cleaning crew can be given the freedom to figure out more efficient ways of organizing their efforts; and as long as it doesn't slow them down or interfere with other employees, why shouldn't they be able to start on the second floor tonight and the fourth floor tomorrow night, so the work is a little less routine and boring?

People list interesting work as the most important item after wages and job security. Only the element of creativity, in its broader meaning, can increase inspiration, energy, enthusiasm, fun, loyalty, engagement, and commitment.

Guidelines for Encouraging Creativity

Act every day on the understanding that each individual needs to believe his or work is important and creative.

Accept the challenge of finding ways to let people in your group find their work interesting. (How high a level in the organization do you have to go before you find people who say their work is interesting?)

Rid your group of dead-end jobs—positions from which there isn't any promotion opportunity or opportunity for additional involvement in the business.

If you are following these guidelines, put one marble into your left hand.

5. Allowing Others to Complete Tasks

A few months before he retires, Tom Brown buys himself a woodworking tool. The big day comes. He gets up the next morning, goes down to the shop, and turns out his first piece, a wooden ashtray. He shows it proudly to his wife, who admires it even though a wooden ashtray sounds like a fire hazard, and besides, neither of them smokes.

She doesn't understand why the stupid ashtray means so much to Tom, and Tom may very well not understand himself. The reason is that in all his years at work, Tom never took anything from beginning all the way to the end.

There's a small tragedy in this for each of us as individuals, but a much larger calamity for the organization. Today too many employees suffer from the gnawing sense that they never get to finish projects—they either hand them off to someone else, just like production-line workers, or the organization reshuffles its plans or budgets, and the project is dropped before completion.

We once met a worker whose job it was to pick up each of the identical parts coming down the belt, inspect it for flaws, wipe it off, and put it back on the belt. No one had told her what the parts were used for—she had no idea. The woman had been doing this same job for 12 years.

Finishing tasks is important to self-esteem. People need to enjoy the feeling of accomplishment.

Guidelines for Allowing Others to Complete Tasks

Recognize that when you deprive others of finishing a project, you disempower them.

Allow people to finish tasks and projects.

> Finishing tasks has both psychological and cultural implications that are hardwired into all human beings. Even children work at play.
>
> A fundamental instinct in all of our colleagues is the desire to finish what they started, and to receive recognition and acknowledgment for their efforts.

If you are following these guidelines, put a marble into your left hand.

6. Encouraging Risk-Taking

One of the challenges within an organization is to encourage risk-taking, while curbing inappropriate risking. This fine line in judgment represents one of the prime distinctions separating manager from leader, and follower from citizen.

Guidelines for Encouraging Risk-Taking

> Recognize and accept that risk-taking is a fundamental element in establishing a mature and fulfilling relationship between each individual and the organization.
>
> Let people see that risk-taking is a welcomed activity—by celebrating successes and perhaps acknowledging (or even celebrating) failures.
>
> Put processes in place to help people take more risks.
>
> Insure that your people understand you encourage risk-taking.
>
> See that each of your people is suitably mentored to curb inappropriate risk-taking.

Shift one marble if you are following most of these guidelines.

7. Setting Policy

When policies are centralized, ownership and responsibility are not in the right hands to create responsible employees, let alone citizens. Those who make the rules own the game. The damage doesn't stand out so clearly on issues internal to the organization, but we can see the damage clearly enough when we look at how one-policy-fits-all hurts our relationships with customers. Organizations that are dominated by companywide policies are less likely to be responsive to customer needs—for example, because of regional and national differences. Centralized purchasing can spell disaster for a clothing store or a national supermarket chain because light-colored sweaters sell better in the South, darker colors in the North; barbecue sauce sells better in Texas and in the South, clam chowder in New England.

Some auto dealers move luxury cars faster, those in the South and Southwest have a continuing demand for convertibles, while dealers in other locales can't keep vans in stock. They don't all move the same product mix. Meanwhile, McDonald's, which built its reputation on uniformity—you know you can always find the same quality and the same items on the menu—now offers regional products (such as lobster-burgers in Maine!).

The marketing arms of most organizations understand the wisdom of sales policies and advertising tailored to each region or locale. The same wisdom needs to be applied throughout the organization, and it's an essential ingredient for getting people to take responsibility: when policy setting is in the right hands, people are much more ready to step up and take responsibility.

Guidelines for Setting Policy

Understand what is meant by "Those who make the rules own the game."

> **Place policy setting in the right hands to create responsible individuals.**
>
> **Establish policies that foster extraordinary opportunity for growth and commitment to the organization's goals.**
>
> **Make your policies aspirational, instead of bureaucratic. (If you're unsure, ask your people.)**

Shift one marble to your left hand if you are following these guidelines.

8. Encouraging Self-Expression

Nobody knows the local business conditions and needs better than the local field team, plant staff, and production-floor worker. Just as the most pleasing words to a customer are "We'll do it your way," the most pleasing to an employee are, "We'll let you do it your way." Whenever Frank Sinatra sang Paul Anka's song "My Way," it always brought down the house—because people resonate to the meaning of self-control, of being able to make their own decisions.

The history of business—and especially the history of technology—is replete with stories of employees who had a product idea that their company turned a deaf ear to, missing out on phenomenal opportunities. When Steve Jobs and Steve Wozniak built their first computer, Wozniak took the idea to his then-employer, Hewlett-Packard; after HP turned him down, the two Steves started Apple Computer—generating hundreds of billions of dollars in revenue that could have flowed into the HP coffers.

Former CEO of Apple Computer, Gil Amelio, had previously run National Semiconductor, where he had instituted a program suggested by management consultant Mike Townsend, of Decision Analysis Corporation. Called "100/100" (because it offered any engineer whose idea was selected up to $100,000 and 100 days to demonstrate feasibility), the program encouraged engineers to come forward with good ideas that might provide the basis for a marketable product—even if the technology was in someone else's backyard, crossing the boundary lines of corporate fiefdoms. Proposals had to be in writing, and had to address how big the marketplace might be and how many of the units the company might reasonably expect to sell. National recognized this wasn't information an engineer could readily put together on his own, so it provided help from people in marketing, manufacturing, and appropriate areas of technology, to aid in developing the plan.

Efforts like this provide ways for employees to satisfy the inherent urge for self-expression. Denying that outlet is one more way companies block their own employees from taking responsibility.

Guidelines for Encouraging Self-Expression

Allow local leadership, team, field staff, and plant-floor employees to act independently, in ways that allow self-expression.

Recognize that "skunkworks" projects, after-hours tinkering, and maverick behavior are all manifestations of frustrated, thwarted self-expression.

Encourage and institutionalize self-expressive behavior *during working hours.*

Shift one marble to your left hand if you are following these guidelines.

Are any of the marbles in your left hand? Most managers and leaders, if they have responded honestly, still have *all* the marbles in their right hand.

SHARING RESPONSIBILITY
AT JOHNSONVILLE SAUSAGE

Can a successful company truly afford to turn responsibility and decision-making over to people at the working levels, and still remain a viable, successful company? Johnsonville Sausage former CEO Ralph Stayer would answer that question with a re-sounding "Yes."

Johnsonville Sausage was offered a very tempting deal by a much larger sausage company—a standing order for such large volumes of product that it would justify building a new, highly efficient plant.

But accepting the order would require hiring and training large numbers of new people quickly. It would mean six- and seven-day workweeks for many Johnsonville people, for months on end, until the new operation was fully up and running. But the capper: the other company would remain free to cancel the order on 30 days' notice—leaving Johnsonville stuck with far too many employees and a large, expensive new plant it could no longer afford.

Not an easy decision. At one time, Stayer would have put his head together with his senior executives and almost certainly decided the risks were too great. But when the issue came up, Johnsonville was already into its new "Let the employees lead" mode. It was the employees who would have to make this work, Stayer reasoned, so it would be the employees who would have to decide. He and his senior managers called an all-hands session and asked the employees to decide: could the company do what would be required . . . and did they want to do it?

Employee teams started meeting immediately to discuss the issues and shape their views. Representatives of each team met in a plantwide coordinating group, and the discussion bounced back and forth several times between the coordinating group and individual teams.

The employees saw only one way to reduce the risk: they would have to increase the already high quality of their product, making it so superior that the other company could find no reason to cancel. What's more, the employees themselves were convinced they could achieve this improvement in quality.

Based on that, the employees decided almost unanimously to take on the new business. For Stayer, "It was one of the proudest moments of my life." And the decision proved sound: the quality did indeed rise, and the other company came back several times to increase the size of its orders.

Impressive indeed. What's the limit of the authority that can be handed over to employees? Stayer writes in his *Harvard Business Review* article that employee teams are redesigning the company's systems and structures. And, he notes, "Right now, teams of Johnsonville members are meeting to discuss next year's capital budget. . . ."

Can you really afford to place responsibility in the hands of employees? "Such thinking has already won converts at the likes of Ford, Goodyear and General Electric," John Greenwald wrote in *Time* magazine ("Is Mr. Nice Guy Back?" January 27, 1992). The article goes on to cite an estimate from Edward Lawler, a management professor at the University of Southern California business school, that more than 80 percent of the Fortune 1,000 firms have at least some degree of employee participation.

So it's not surprising to find that Conrail gives employees authority to assemble problem-solving teams when they see a need. When Stanley Gault took over as chairman at troubled Goodyear, he installed a similar approach and later proudly told the *Time* reporter, "The teams at Goodyear are now telling the boss [Gault himself] how to run things. And I must say, I'm not doing a half-bad job because of it." Mobil Oil now gives field teams the responsibility to decide when and where to drill—decisions once reserved for management.

But it's not an easy transition: onetime Ford chairman Don Petersen found a lot of people giving "lip service" to the sharing of responsibility, but complained that "The moment they find themselves in difficulty, they revert to form."

WRAP UP

The eight keys point people toward taking responsibility, and help employees recognize they are the rightful owners of responsibility within the organization.

The Eight Keys to Responsibility-Taking empower employees to:

1. Plan.
2. Set priorities.
3. Remove roadblocks.
4. Encourage creativity.
5. Allow others to complete tasks.
6. Encourage risk-taking.
7. Set policy.
8. Encourage self-expression.

In a command-and-control environment, these eight are the duties that *managers* hold, and employees are rarely allowed to tackle. In the Shared Values environment, the responsibilities shift . . . and everyone gains.

11

Thank Socrates: A New Approach to Decision-Making

If you come to a fork in the road, take it.

Advice from baseball great Yogi Berra,
renowned for his garbled logic

H ow do people make decisions, and on what basis are problems solved? Too often, the important decisions in business and government—and in our personal lives, as well—seem to be based on precepts formulated at the Yogi Berra level of wisdom. Too often, solutions are proposed before anyone has had a chance to arrive at a clear understanding of the problem, before any information has been gathered. We've lost the art of tackling a problem by first gathering pertinent, current, valid information that could show the way toward the most appropriate solution. Most of us function on automatic pilot, rushing headlong toward a fast solution without analytical time to evaluate the situation. The answer seems obvious, and

so we make a decision from a seat-of-the-pants emotional response and leap at a distinctly wrong solution.

The result: far too many problems wearing Band-Aids—
problems covered over rather than cured. And these short-term
patches mean that the same problems keep reappearing over
and over. Instead, managers need to initiate and teach a new
problem-solving process.

There is a precision process to problem solving that some
perceive as an art form. Once you learn how to make decisions
and find solutions, when to use your intuition and when to
search for more information, and how to measure the success
of a decision, it will become an artistic skill that has the precision of a Rembrandt painting.

To become really powerful at making decisions, you must
stay open to the possibility of redefining the problem even as
you begin the first steps toward a solution. The initial search for
facts takes you forward but may also send you backward as
more information impacts your understanding of the situation;
you must be strong enough to move forward and backward at
the same time.

Problems that need to be solved should be treated as prag-
matically as e-mail that needs to be answered, and should have
a target date for the solution. Until a manager has an action list
of problems waiting to be solved, with target dates for solutions
attached to each, the problems are just amorphous worries
causing stress. A problem that isn't being addressed leads to
worrying the problem, not *solving* it. Managers need to view
their productivity in terms of problems solved as well as in
terms of deadlines met or products shipped. If you are not
scheduling thinking time, you may not have the appropriate respect for the importance of problem solving.

Managers also have the responsibility for developing within
their direct reports the ability to make decisions based not only
on intuition, but also on facts and data. This demands a solid
decision-making process that the manager can model—a

process that insures a clear statement of the problem and measures the success of the solution.

VALUES-BASED DECISION-MAKING

Over the centuries a classic model for wise and powerful decision-making has been the Socratic method. Based on this model, which still retains its potency, we've identified five steps to *values*-based decision-making—easy to follow through, easy to role model, and easy to teach others:

> *Step One:*   Gather information and set priorities.
> *Step Two:*   Identify the problem, focus on the issue.
> *Step Three:* Plan the solution.
> *Step Four:*   Implement the change and work the plan.
> *Step Five:*   Measure the results.

Step 1. Gather Information and Set Priorities

It's a human trait to hang labels on things and people, and then, having identified a category, to make assumptions based on our past experience. Foreigner, Republican, suburbanite, commuter, manager, white-collar worker, Frenchman, parent, homeowner. . . .

The process makes good sense—it saves the ordeal of starting from scratch to evaluate every new person and situation that comes along.

Five years ago the child who didn't pay attention in class was looked on as misbehaving and a troublemaker, but now is likely to be classified as having Attention Deficit Disorder. Once we have a label to apply, we start looking at the problem differently, and our choices expand or become limited because of it.

The trouble comes when we apply the slap-a-name-on-it routine to problems in the workplace. "Customer didn't get the shipment? Another shipping delay; happens all the time. See that the order gets sent right away. What else is happening?" It's just another shipping delay, so we'll do what we always do with shipping delays.

Naming the problem often short-circuits any sense of a need to search for the true problem, issue, or cause. Any difficulty that happens frequently enough to have a label is almost certainly a problem you should be addressing.

Once you've recognized a recurring situation, gathering information must come before anything else. Start by putting together a cross-functional team. And look on the effort as a *discovery* process, rather than the kind of problem-solving program that goes around looking to pin blame on people.

Here's another requirement that may surprise you: include some people who know little about the problem or subject matter. These people, in asking questions to understand the issues, frequently bring things to light that would otherwise have been overlooked.

If you don't have time to do it right, there's little point attempting the problem-solving process described here; recurring problems (as opposed to emergencies) can't be solved under a time constraint.

Don't focus on a narrow issue; instead, set out to investigate in broader terms and solve the underlying problem. So in stating the goal you might ask, for example, not "How do we solve this shipping problem?" but rather the more general "How do orders get delayed?" The difference may seem minor; it isn't—a narrowly worded goal will limit the scope of the team's work and probably not lead to a solution to the ongoing problem. The wording can be critical.

Guidelines for Gathering Information and Setting Priorities

Use 30 to 50 percent of the available time for gathering information—we never dispute or mistrust information that we ourselves have gathered.

> Allow and encourage *all* members to actively par-
> ticipate in the discovery process. This step will
> include discussion, disagreement, and the recon-
> ciliation of opinions and interpretations.
>
> Support, enable, and empower members to es-
> tablish priorities centered on gathering informa-
> tion, not on attaching labels.
>
> Once enough information is on hand, and priori-
> ties, issues, and benchmarks have become clear,
> identify and narrow the issues to look at. *(One of
> the challenges will be to focus the investigation.)*

Step 2. Identify the Problem, Focus on the Issue

The process of identifying the problem frequently leads the
group to encounter a much broader spectrum of challenges
than the members have time or resources to handle. Hence the
issue becomes one of trying to narrow the focus. For example, a
group has tackled the problem of customers often getting shipments late; the information-gathering process may turn up factors contributing to the problem in shipping, sales, purchasing,
manufacturing, and other areas.

While in Step 1 you asked, "What do we need to know?" in
Step 2 you focus on bringing the scope of the effort down to
manageable size. At the same time you need to begin thinking
about how you will measure the results: you will have no way of
establishing success—to your own group or to others—unless
you are able to measure the results, in terms of changes in
processes, job assignments, policies, or procedures.

Guidelines for Identifying and Focusing

> Identify the problem based on objectively gath-
> ered facts.
>
> Use a consensus approach (Chapter 7) to reduce
> pushback and eliminate sabotage.
>
> Consider how you are going to measure results.

Step 3. Plan the Solution

Planning the solution to any problem, issue, or reachable goal requires assumptions about the team or group's ability, commitment, and perseverance. Instead of the team leader driving the process, the leader prepares the group members to take over and drive the process themselves.

Guidelines for Planning the Solution

Involve all members in defining the steps of the solution. *(Holdouts will ultimately sabotage the process.)*

Empower members to establish sub-task forces if necessary, knowing that some subjects are just too broad.

As leader, begin changing to an advisory role— suggesting and probing for clarity, becoming a resource, but allowing the group to take over the planning.

Step 4. Implement the Change and Work the Plan

Plans must be implemented by the same core team of people who have gathered the information, identified the problem, and planned the solution. But implementation may require expanding the team.

It's tempting for a leader to drive the implementation—this is clearly more efficient in the short term. To do so, however, would forfeit the benefit that team members get from doing this themselves. Team participation is a vital learning process; as leader, you need to confine yourself to the role of "coach" (as we'll see in the next chapter).

Guidelines for Implementing the Change and Working the Plan

The people who implement the change must be the same people who gathered the information, defined the problem, and worked the plan.

Insure that the group develops a means of measurement.

Since few plans are perfect the first time, encourage members to be proactive in making adjustments. Take nothing for granted; remain committed to continuous improvement.

As leader, employ a coaching style that corrects, models, and praises evenhandedly. Let the team make its own decisions.

Step 5. Measure the Results

In a problem-solving effort, people feel empowered when they are allowed to measure their own results. And the more often they measure, the more energized they become. Removing this energizing activity from the group and giving it to others would signal a loss of trust. The activity reconfirms engagement in the plan, and reconfirms the strategy.

Often one or two people will volunteer to take on the job of measuring. Resist this if you can—let everyone share the heady firsthand experience of obtaining results of the planning effort by encouraging everyone to take part—sharing the task on a rotating basis.

Finally, celebrate the efforts and the results.

Guidelines for Measuring the Results

Develop benchmarks and goals.

Measure; allow and encourage all members to actively participate in the measurement process.

Insure that the people who are doing the work routinely report the results.

Celebrate large and small successes.

WRAP UP

There are five steps of the Values-Based Decision-Making process for a group or team to use in resolving a problem:

1. Gather information and set priorities.
2. Identify the problem, focus on the issue.
3. Plan the solution.
4. Implement the change and work the plan.
5. Measure the results.

Remember that a major value of the process lies in the experience that group or team members gain—experience that prepares people for assuming higher levels of responsibility within the organization.

12

From Manager to Coach, and Coach to Wise Counsel

You must manage as if you need your employees more than they need you.

Peter Drucker

Most companies are over-managed, and most employees need less instruction than we think.

Jim Treybig, then president of
Tandem Computer

THE TRADITIONAL MANAGER

What are the most profound changes taking place in business today? Reengineering. Down sizing. Layoffs. Outsourcing.

And whose role in the company is the most affected? Sure—it's the manager. The role for managers is changing.

Nothing beats the energy of a start-up, where nobody has had time to write down a lot of rules, and it's all, "Wow, great idea, let's go do it." But for a traditional company, even for the person sitting in the CEO's chair, it seems impossible to tear down the barriers. For middle managers, it seems futile even to try.

The traditional manager—the manager in a standard command-and-control organization—has assumed a role of deviser of goals, planner of tasks and strategies, protector of the group's sanctity, limiter of risk, maker of standards and policies, and director of action.

Reinforced by promotion ladders and incentive programs, these roles have defined the manager's raison d'être, his or her reason for being. And the manager who can carry out the tasks at the highest levels of expertise is described as a good leader. In most traditional organizations, there are few good leaders. Good managers, yes, but not many true leaders.

Leaders require followers, and in the traditional organization, good followers were seen as reliable, responsive, and willing to march loyally in whatever direction the leader pointed. The followers' willingness to be compliant in exchange for organizational loyalty, the keystone of the command-and-control organization, was the glue holding the company together.

During the late 1980s and early 1990s, the organizational compact that had stood as the hallmark of this fundamental relationship between the employee and the leader disappeared from many companies—particularly those that held paternalism as a core value. Without fanfare, the traditional organizational compact eroded, leaving a void that has not yet been filled

The Traditional Manager in Action

Traditionally, management has assumed the role of director of action—knowledge keepers and knowledge providers. This traditional role has in many cases defined the manager and defined the manager's subordinates. With these roles clearly defined, a line is drawn on the ground dividing the rights and privileges of the managers from those of the workers.

Hierarchical organizations define leaders by the number of followers they have, and by their ability to "get the job done" while staying within the confines of the organization's policies and procedures. Managers were gauged by two criteria: whether the group reached its goal or quota; and whether the manager demonstrated strength and talent in tasks such as:

Planning
Priority setting
Removing roadblocks
Establishing creative work strategies
Retaining integrity of the group while minimizing risks to
    the operation
Supporting and interpreting existing policies

Our ideas about managers have been as rigidly fixed as the ideas at military schools like Virginia Military Institute and the Citadel about the acceptance of women into the student body.

At the same time, the employees, as well, have been programmed to understand their passive and compliant role in meeting the demands of their supervisors. Both the managers' and the employees' roles were conveniently packaged—employees gave comfort to managers, and in a way, managers gave comfort to employees. The compact between them assured continued employment in exchange for continued compliance and loyalty.

Today, as that relationship crumbles, employees are becoming increasingly passionate about a loyalty that traditional companies no longer find themselves in a position to promise. Unions in some of the recent auto union negotiations have begun asking for guarantees of long-term employment. Wage increases are taking a back seat to job-security demands. And the database we maintain on employee attitudes tells us that security is the number-one issue in American business—for workers, and for managers as well.

Just as organizations seek new ways to define themselves, so also a different basis needs to be found for the relationship between manager and worker. Business can no longer afford to encourage compliant, timid employees who are afraid to reinvent their jobs. Partnership, not paternalism, will be the objective of successful organizations in the future.

CHANGING THE RULES, CHANGING THE CONTEXT

The only way to induce change is to change the rules, by changing the context. But we can't ask people to somehow "come to it"—we have to help them discover it. The reason: the change is not obvious. Tell a manager to stop managing in a traditional way, and he has no idea what he would replace it with. Tell employees to stop being compliant, instead take more risks, be more assertive, be proactive—without furnishing a structure for their migration—and they will be stopped in their tracks.

The best way, we've found, for introducing new ideas about manager/employee roles is to invite both groups to experiment with new skills and tools, and a new philosophy, centered around a real-world task in their own everyday work environment. The effort should be at the same time a useful project and a revealing experiment in changing the context of the workplace.

MANAGER TO COACH

We rename the manager as the "process leader"—asking the manager to become the teacher of a new process of decision-making. A manager understands where he or she is today; that he has to experiment with becoming a teacher of a new decision-making process; and that somehow, some way, those eight marbles—the eight Responsibility-Taking elements—have to be given back to the employees.

So there will be three dynamics all going on at the same time:

Managers transforming from managing people to teaching processes;

Employees coming to understand that they have to embrace as their own the eight elements of Responsibility-Taking; and

Managers learning to accept that once the effort is set in motion, they will never again be driving people but will be setting the conditions for their embracing the new model of Responsibility-Taking.

HOW IS A MANAGER LIKE A COACH?

How is a manager like a coach?

Sounds like the setup for a joke. "I don't know—how *is* a manager like a coach?" Today, managers are not like coaches. But tomorrow they will be. That's the transformation we need to make—turning managers into coaches.

What's so wrong with being a manager, that we have to change hats? Remember—the new manager, the Shared Values manager, is going to give up responsibility to his people. The manager doesn't have to micromanage any more, so will have time for other things, like being a coach. But what do we mean by the manager as coach?

Consider what the coach does for a sports team: picks the players, assigns the positions, provides the strategy, directs the training, critiques the performance, and makes sure the players have the equipment they need.

The Shared Values coach in business does almost all the same things, but lets the people lead. She makes sure everybody knows the strategy, works with her people on who takes what assignments, sees that everybody has the training needed, and makes sure they have the resources and tools to get the job done. She doesn't direct these activities, but helps facilitate the awareness that these tasks need to be done. And when somebody's game isn't going according to their game plan, she's on hand to support, advise, and guide. Coaches encourage the other members of the team to help their comrade.

If you've been active in a sport, especially a team sport, you've probably had some good coaches who helped you improve your performance. But you may never have had a coach you'd want to have for a boss at work, day in and day out.

On the athletic field, in the gym, at the swimming pool, the coach can shout, discipline, punish—and still have the respect and admiration of his athletes. In business, the coach has to depend on his ability to *influence*. A heavy-handed coaching style just doesn't make it in a Shared Values environment.

A man we knew who was a member of the Ski Patrol at Sun Valley some time ago began little by little accepting students on the side, and the experience of teaching began to change him profoundly. This was a young man who could talk about ski wax and chairlifts all day, and Lake Tahoe versus Mont Blanc. He read every ski magazine but never opened a book, and probably hadn't had a single idea in his head since he left high school.

But as he grew into being a really good teacher, he was taking what he knew intuitively and struggling to get people to understand. At first it was all "Put your weight here, put your pole there." But the little nitty details began to coalesce into something else. A man who probably couldn't spell the word was developing a *philosophy* about his subject, changing into a person with ideas and understanding.

Does all that have anything to do with the world of business? You bet it does. The journey from manager to process leader to coach turns out to be the very same journey of *self-discovery:* a discovery of skills, competencies, and philosophy.

Not every great athlete can become a great instructor, of course. Many of us have taken golf or tennis lessons from one-time touring pros who seem to have *the* answer for our game, and then the next time we go out to play, we're right back where we were before—the training was good at the moment but it didn't stick, it didn't produce any lasting change in our behavior.

The transformation to coach demands that you desire to embark on the most challenging event of your life as a manager, in which you aim to become an *ex*-manager. At that point you don't have to manage any more; instead you've become a teacher, mentor, enabler—a *leader* of other people.

The reason not everyone can make the cut is because not everyone is willing to root for the success of others. Those managers who are only interested in their own personal careers will never succeed as coaches. And they'll never be able to establish a Shared Values environment.

TRANSFORMING MANAGERS INTO COACHES

If you have the commitment, the following four steps to becoming a coach will work anywhere, in any industry, in any organization, with any kind of corporate culture.

1. Learn to *partner* with your people.

    Though your employees may not do things exactly the way you would have done them, the work *will* be accomplished.

    That sounds like it requires a truckload of trust—but it's a two-way trust. Managers have to trust their people, but they also have to establish a fair environment where the people also trust them.

    In each direction, the burden is on the manager. This is not something that will ever get started if you sit behind your desk and wait for your people to come to you carrying a flag labeled *Trust,* and ask you to lead the parade toward a new horizon. But trust is one of the Shared Values, and if you have a Shared Values philosophy underway first, your employees are already set up for this.

2. Set an example that will encourage your employees to allow you to coach them.

    Given a free choice, most employees would naturally prefer to maintain the same old comfortable relationship with their manager. Employees have to understand the change in process and become ready to embrace it.

3. Set an example that will encourage your employees to be mentored by everyone in the organization.

    Recall that this is one of the eight Shared Values.

4. Be willing to let people experiment on their own, with new approaches and new ideas.

    What do you do when somebody comes to you with a great project he's dreamed up? He's written a plan showing it could be done with a reasonable effort, and could produce very good results for the company. What do you tell him?

    Most managers say something like, "If it's not in my business plan, how can I?" They can't drop everything just because somebody has what looks like a good idea.

    But what happens to your people if every time somebody comes up with a new idea, you say, "We can't do it, we can't even try it"? Of course: they stop bothering to come up with new ideas.

LEARNING TO BE A COACH

Co-author Bill Simon holds a commercial pilot's license for fixed-wing aircraft and helicopters, with instrument rating. Earning those ratings took a lot of hours in the air with flight instructors, and Bill remembers the frustration of sessions that involved a flight out to an aerobatics practice area, a detailed explanation of a maneuver by the instructor, followed by several demonstrations by the instructor . . . and then there was hardly any time left to practice the maneuver before starting back to the airport. These teachers enjoyed flying the chandelles, lazy-8s, 8s-on-pylons, and autorotation landings so much that they had to be reminded the student didn't learn just by watching, that real learning doesn't start until you begin trying the new technique yourself.

It's like a bachelor who marries a woman with children. Nobody teaches him to be a father but, finding himself in the role, he discovers the way. If he had taken a course in parenting, he would have gained information and knowledge, but wouldn't necessarily have become a fit father.

In the same way, managers learn their new roles by doing— by introducing the Shared Values philosophy to their people. And the people, in turn, learn their new roles not by listening to the manager, but by getting in there and putting the ideas to work themselves. Sitting in a classroom has its place, but it's not where the real learning happens. The best experience, as always, is hands-on experience.

In this process, we don't try to "transform" people, we don't "fix" them. Managers learn by experimenting with the transformation themselves, and their people learn by also being given the chance to experiment.

Just as there are many different styles of managing, so there are many styles of coaching. Some do it mostly by leading, some mostly by demonstrating examples, some mostly by personal involvement, some mostly by inspiration and motivation.

One manager will develop a coaching style that's very detached and non-emotional; another will make it very personal. Each person needs to find the approach most comfortable for them.

BECOMING A "WISE COUNSEL"

There is another, higher level beyond coach: what we term the Wise Counsel.

Unlike a manager, the Wise Counsel doesn't supervise, or establish goals or priorities. Like a coach, the Wise Counsel teaches, but at a higher level. He or she is a resource—sharing the benefit of experience, counseling by bouncing ideas back and forth. He or she is an advisor, *who has no responsibility for the traditional result or success of the project.* The Wise Counsel's greatest role is to insure the success of the *people.*

So if responsibility for the results of your group belongs to you, you are still functioning as a manager; if it is shared between you and the group, you have succeeded in changing the relationship, turning yourself into a coach; and if responsibility for the results actually rests with the group—congratulations! You've become a Wise Counsel.

It's similar to the pattern of changes in relationships of loving parents as their children grow up: With a baby, the parents provide protection, caring, and a response to fill all needs. With a teenager, the relationship shifts gear, toward less control and some measure of autonomy—the child is taking more responsibility for her actions and behavior. When the youngster turns into an adult, the responsibility shifts much more away from the parents—leaving them in a position to offer guidance and support, but (usually) free of any need by the child for help in accepting responsibility and making decisions.

A Wise Counsel is a manager who has evolved into a key resource for the group members. Besides coaching and teaching them, she's able to spend time communicating the values and standards of the company. She spends the bulk of the time thinking strategically about the role, purpose, and job of each person in her group. She serves as a reality check to individuals and groups, and adds a continuity to all the processes that the organization wants to try.

When an organization pursuing the Shared Values philosophy reaches this third stage, executives hold *groups* responsible for results, not the managers of the groups. That helps greatly in the managers' transformation because this change is a contextual one.

And if you're giving the individual employees that much lati-

tude, you're going to end up with a very free-spirited organization, one that embraces universal education and
experimentation. It's an organization that by its nature is enormously flexible, forgiving, and dynamic.

CAVEATS

Some managers can't bring themselves to give up the "traffic
cop" role, can't (or won't) give up being in control and setting
policy. They are so fixed in their ways that they are not likely to
change.

Another kind of fallout comes from managers who are unable to understand the role they are carrying out—people who
are more interested in who they are, and how many people
report to them, and how much longer before they get a corner
office.

The question becomes, Are you yourself able to put aside
the old values and start changing into the kind of effective manager who wants to be judged on his mentoring skills, and the
ability of his people to stand on their own feet and think as individuals—turning people into entrepreneurs?

WRAP UP

Becoming a coach and then a Wise Counsel doesn't require
some skill or ability you have to be born with—as in the joke
about a man who arrived at Grand Central Station in New York,
ran out to the street, hailed a cab, and breathlessly said to the
driver, "Do you know how to get to Carnegie Hall?" In a heavy
Bronx accent, the driver answered, "Practice, practice, practice!"

That's the way you transform yourself from manager to
coach, and then to Wise Counsel: practice, practice, practice.

You'll likely find many of the new ideas and approaches of
the Shared Values environment uncomfortable for you and your

group when you first try to put them into use. Think back to the first few hours you spent using a computer, and contrast that to the ease with which you use your PC today. The Shared Values Operating System is designed to give everyone time to consider alternate approaches—to discover new ways of getting the job done by partnering with people. Your new role is to create the conditions for everyone's success so they can play at the top of their game.

And how about the challenge if you work—as you probably do—in a command-and-control environment? In fact, every person who has others reporting to him can break down the hierarchical structure in his own part of the company. Everybody can. Fix your own operation, and other people will start coming around to find out what you did.

Imagine the satisfaction as you work quietly to establish a new context centered around people and find that your work group is indeed beginning to function with a willingness to take responsibility, leading to a true entrepreneurial spirit.

"People want to be great," wrote Ralph Stayer. "And if they're not, it's management's fault." These words should echo through the management corridors of every organization.

FRAMING YOUR SYSTEMS TO PEOPLE NEEDS AND CUSTOMER NEEDS

People Systems

So far we've been dealing with the values and principles that underlie the Shared Values philosophy.

Now we come to the practical aspects: putting Shared Values to work in your organization. The techniques presented here are those we have evolved through our years of installing Shared Values in companies internationally, the process is introduced in five phases, or "modules" (see the rollout chart below).

A few of these topics have been dealt with earlier; most are covered in the chapters that follow.

A People Operating System™

The Unifying Force

Enhances Training and Transformation Initiatives

FIRST GENERATION
Learn New Skills
- New Skills & Competencies
- New Subjects of Knowledge
- Adapting to a Learning Organization
- Sales Training
- Technical Knowledge
- Customer Service Focus

BUSINESS EXCELLENCE AWARDS
Retain Acheivement Levels
- Baldridge Criteria
- ISO/QS 9000
- CABE Criteria
- Presidential Award for Quality

SHARED VALUES PROCESS® OPERATING SYSTEM

Values Based Context

Social Psychology Models
- SHARED VALUES Balanced Business Values™ & People Values™
- STANDARDS Guidelines for Appropriate Conduct & Actions

Organizational Development
- PEOPLE SYSTEMS™ Hiring, Orientation, Compensation, Mentoring, Career Development, Citizenship, Quality & Communications
- REDESIGN STRUCTURES & PROCESSES Customer-Focused Process Groups

SECOND GENERATION
Maximize Human Potential
- Empowerment
- Positive Thinking
- Habits of Personal Character
- Visualization
- Affirmation
- Incentive & Motivation
- Aversion Therapy
- Positive Reinforcement
- Sensitivity / Diversity Training
- Paradigm Shifting

QUALITY BUSINESS PRACTICES
Enhance Retention
- Open Book Management
- Process Engineering
- Core Competencies
- TQM/CQI
- Activity Based Management

Rollout chart: the sequence of introducing the modules of the Shared Values Process Operating System into an organization.

13

Redesigning People Systems to Support the Shared Values Philosophy

In the days of the Cold War, a Russian and an American were fishing on opposite banks of the river which at that place divided East from West Germany. The Russian had no success while the American landed several fish. Eventually the Russian shouted over in his broken English, "How do you catch so many?" To which the American replied, "Over here the fish are not afraid to open their mouths."

Anonymous

WHAT ARE PEOPLE SYSTEMS?

An Operating System is in large measure the culture or context upon which everything revolves. In the same way that every organization already has an Operating System, so every organization also has a series of what we refer to as "People Systems," a series of eight practices that define the ways in which the organization relates to its employees, and the way employees relate to each other. They describe the obligations of the organization to its people, and those things the organization needs to do to help its employees perform at the top of their game.

And just as it takes unremitting pressure from corporate leadership to recognize the organization's Operating System and conform it to the belief systems and what the company intends to stand for, so the People Systems need to be brought into the open, examined, and made over into one single coordinated system that reflects the goals and intentions of the company. Then the system needs to be taught to and adopted by all employees—a procedure in which the employees themselves play a major role.

People Systems are in many cases below the surface of what is realized or appreciated by the employees and the customers. Often the existing systems don't reflect what we would like to deliver to our staff or to our customers, yet they drive everything about how the organization operates and behaves.

THE EIGHT PEOPLE SYSTEMS

Through understanding the People Systems, an organization takes the first step toward being able to bring awareness. In time, with continued effort, the company molds all eight systems until they reflect the values that the organization stands for. (The techniques you'll use to carry this out—forming a separate task force to address each of the eight systems—are detailed in the next chapter.) Once shaped, appropriate interactions with all other parts of the Operating System are possible because there is a balance between Business Values, People Values, and People Systems.

Hiring

A company defines itself, intentionally or unconsciously, by its hiring criteria. The people you choose to hire will in some ways adapt to the environment of your organization; but gradually, over time, enough good hiring choices or enough bad ones will begin to influence and change the context of the organization, strengthening how well it functions or undermining its ability to serve its customers and meet its goals. Everything is driven by our hiring criteria.

A large farm cooperative operating in several states had spent a six-figure sum to hire a new national sales vice president, bringing him in from another part of the country and paying his resettling expenses.

A few months later, a group of company people were in training for adopting Shared Values in their company and the facilitator was discussing the issue of resolving conflicts.

"What would you do if you couldn't go directly to the person creating the conflict?" one woman challenged the facilitator.

"You always can," the facilitator answered. "You're not going to fight or make a complaint, you're just making a request— you can't get in trouble."

With a little heat rising, the woman disagreed.

The facilitator asked, "You'd fear for your job just because you made a request?"

"Yes."

"Then you have to level the playing field," the facilitator advised. "Get the help of a manager who isn't afraid of the person."

"What if you know the person who's causing the problem will smile and agree, but later will find a way to get you fired?"

By now it was clear that everybody in the group knew who the woman was talking about. Two of the other women looked as if they were about to start crying, and the men were squirming while trying not to look timid.

The facilitator determined they were talking about a high-level executive, and advised that they needed to take the matter directly to the president.

The president was alarmed at the possibility that his expensive new VP was causing problems, but it didn't take

long to confirm that he had a terrorist loose in the
organization. He recognized that the damage being done far
outweighed the dollar considerations, and, after much
discussion with the VP but no sign that he understood the
problem and could improve, he asked the man for his
resignation.

It's all too common for a company to hire a new manager or
executive, and then discover that the person doesn't fit well
into the culture of the company. How do you avoid this damaging situation?

First step: get rid of the traditional, off-the-rack hiring package. In its place, set up a cross-functional task force and charge
it with the job of identifying the basic attributes of the kind of
person your organization wants to attract.

You want to establish an across-the-board series of attributes that people throughout the organization will agree are
important for all jobs, regardless of if they are line or staff, factory or office, manager or worker. Examples:

- A strong work ethic
  - How important is getting to work on time?
  - Keeping members of the work group advised that a
    project is slipping?
  - Being trustworthy?
- Experience and qualifications
- Individual and collective responsibility
- Life attitudes
  - Friendliness, cheerfulness
- Attitudes about learning and continuous improvement
- Motivation and validation issues
- Opinions on diversity

. . . And of course there are countless other possibilities.

Gil Amelio, the former-CEO of Apple Computer, has a personal list that appears in his book of business principles, *Profit from Experience*; the items are listed in what Amelio considers to be their order of importance. Note that values is first on the list, and experience—which, he points out, is where most companies place the strongest weight, ranks down in third place. The Amelio list:

- Strong personal/ethical qualities
- Brightness
- Experience
- Sense of humor
- Ability to work hard
- Intuition

Once your organization has endorsed a new set of criteria, you need to take a broom to your hiring system. As mentioned earlier, one reason for the great success of Microsoft lies in its direct approach to hiring: Human Resources is only a coordinator of the process. With hiring decisions left to the group, candidates are interviewed by the people who would be their closest contacts in the company—the group of peers and their leader. These are the best people to judge which of the prospects is on the right wavelength, and which most closely matches the profile of attributes the group has defined.

An additional procedure we recommend: a "process leader" should be assigned (in some cases the immediate superior, but not necessarily), who will coordinate the process of interviewing and evaluating. During the interview process, each person who meets with the prospect should be required to record, in writing, their reasons for recommending the person be hired, or reasons against, and then must sign their names—a step that leads people to consider their decision more carefully.

The process doesn't end with hiring. You also need to go through the same rigorous attack on policies about retention and termination. To what lengths will the company go in order to retain employees? What constitutes a fair, clearly understandable termination policy? (In most companies, if you asked 20 employees "What would get you terminated?" you'd find this an issue that gets you 20 different answers.)

Hiring is the pivotal issue in your People Systems because it establishes the baseline for behavior and expectations. Once you've worked out the criteria for hiring, you're ready to move ahead on the other systems.

New Member Orientation

Some companies have found that calling every employee a "member" brings a distinct change in attitude. For a new employee in particular, the implication is the same as joining a club. Following this idea, we use the term "new member" here for people newly hired.

What is the defining moment in a new member's work life with your organization? Answer: the very first moment they walk through the door. One organization that took this seriously has changed forever the new member experience, and the career course of these new people has led to some amazing success stories.

In a typical organization, the enthusiastic new member comes to her new office or factory a little bit early the first day. What will she find? Who will be waiting for her? Will she see her name in lights? Will her work group be assembled to welcome her—maybe even in some kind of party atmosphere? Or will she meet with disinterest, paperwork to fill out, and lots of getting lost and asking directions? Are new members treated as honored guests, or the low people on the totem pole? In terms of investment dollars, how much of the money we spend on hiring the right people do we waste by pouring cold water on their enthusiasm on that very first day?

One organization we know set out to turn the first-day experience into a visit to *the best operation in the world.* Considerable effort went into this—and we all know that you never have a second chance to create a first impression.

Some companies assign a "buddy" to accompany the new member, for the first day at least, sometimes longer. Others put the paperwork off to the second day or later. A complete tour of the facility is always a good idea. Many great people-companies, including Disney, Alaska Airlines, Harley-Davidson, Ben & Jerry's, Pepsi-Cola, and Westin Hotels have gathered a group of stories reflecting the organization's success and pride that is told to all new employees.

Getting a great person off to a great start is a win for everyone.

Compensation, Incentives, and Motivation

While there are, if anything, too many books on these subjects, some of the values-based aspects seem to fall through the cracks.

One powerful carrot for encouraging workers and managers alike to shoulder greater responsibility lies in creating a compensation formula that equates earning with the acceptance of *added* responsibility. One example was mentioned earlier: the sausage company that doesn't provide longevity pay raises, where workers can increase their pay only by learning new tasks and skills.

Generally, the more elements a compensation program has (short of becoming so complicated no one can understand it), the more sophisticated it is and the more impact the motivation has on the individual—allowing the organization to reinforce desired skills. Does the program at your company provide pay increases based on factors like experience, technical knowledge, profit sharing, and personal incentives?

Companies such as GE and 3M have designed compensation programs that embrace many aspects of a person's work effort. But it's not just for large organizations. Century Circuits & Electronics has developed a creative approach to establishing pay levels with the partnership of their union representatives, using a two-dimensional matrix based on core competencies and attainment levels. Another small multinational operation, Ault, Inc., has established a creative approach that combines core competencies with cross-training. Ultimately you must ask the question: is our pay plan in line with the other People Systems issues? If not, what are you going to do to change it?

Personal Improvement, Coaching/Mentoring, and Feedback

This encompasses what in the past would have been called "appraisal"—a process we discourage and that we argue should be dropped from corporate practice. Appraisals influence behavior, but there's no research to suggest that they influence outcomes.

In Alfie Cohn's insightful book, *Punished by Rewards,* he convincingly argues the view that when you pay children to read books, you devalue the reading. The joy of the experience, the pleasure of books, which you hoped they would discover, is lost when reading becomes instead merely a means to buy CDs or computer games. How sad for young people to have this pleasure stolen from them forever. In the same way, much research in the business world supports the notion that compensation should not be linked to a person's appraisal—an approach that undermines the value of the information the employee gains from the process.

Deming and others have railed for years against the annual or semiannual performance review, especially when linked to compensation. But what else can an organization do?

Appraisals are a poor way to attempt to change people's behavior: behavior won't change until you change the context. Threaten, incentivize, do what you like, and people will not change. Consider this: if you doubled the pay of every person in your operation, would they become more efficient, more responsible, more creative, more trusting, more talented? Of course not.

Rather, why not allow each individual or group to establish measurements? Replacing appraisals with a personal capabilities review, where the aim is not to appraise but rather to develop, offers the best alternative approach for this contextual change.

In a personal capabilities type of review, the individual is asked questions centered around developing individual skills. Coaching, modeling, mentoring, and setting developmental skill goals should be the approach to take. Because each person is an individual, even though he or she may do a job others do, each requires a personal approach. Establishing an individualized program for each person may at the onset require more time, but in the long run, developing people is a key to success for any organization that wants its people to play at the top of their game. Creating the conditions for success for each individual starts with helping each individual grow. Now you have an approach that dovetails well with an active mentoring program.

Feedback systems work well to enhance this review process—either self-monitored by the individual, or monitored by a third party (i.e., someone other than the person's supervisor). We have found this monitoring process, common in Japanese companies, reduces in-house politics. Feedback can be very tricky; many people have difficulty providing feedback effectively to another person. But everyone benefits from having an uninvolved group member or supervisor mentor them unselfishly.

Long-Term Career Development

Long-term career planning for employees simply isn't part of the thinking of many organizations today—one reason some companies lose so many valuable people.

Because the young person joining the workforce today is likely to have five or six different occupations over his or her working life, there's a compelling need for each individual to have a long-term learning program—which should be developed jointly by the individual and the organization.

Part of the secret of good hiring is finding people who want to grow; part of the secret of keeping employees mentally awake and adaptable as the organization's needs shift lies in nurturing the desire for lifelong learning. Long-term career development is a philosophy based on learning, but it's built on a foundation of how the organization "feels" about people, how the organization is willing to treat, nurture, and grow its people.

The organization itself needs a clear philosophy centered around what it understands as its responsibility toward its employees, including the extent of the investment it will make for the long-term development of each employee. In the best of all possible worlds, the result is a two-way agreement: the individual accepts the responsibility to continually grow, and the organization accepts the responsibility to continually reinvest in the individual's growth.

Different organizations come up with different approaches to the "posteducational experience." A supervisor at a bank wanted to go back to school for a music degree; against expectations, the company agreed to pay, because it believed that *any* further education would enhance an employee's job performance. And, indeed, the employee became less one-dimensional, in the process gaining self-confidence, even to the point of starting a company dance band. An employee who had been intractable and hard to reach became more flexible through the experience of education, more open to different points of view.

Taken together, Personal Improvement and Long-Term Career Development provide the one-two punch of short-term and long-term approaches to future survival and success. The combination signals to each individual his importance to the organization, and emphasizes his personal responsibility to improve skills and competencies.

Do these approaches make the traditional process of appraisals unnecessary? Almost all organizations that have adopted the People Systems agree that they do.

Communications: Information Transfer

We are, if anything, overloaded with communications. Everyone has heard the horror stories, if not experienced them, of coming back from a business trip to find a hundred, or several hundred, e-mail messages stacked up electronically—most of them from people waiting for a response.

Companies these days give ample attention to the search for ways to make electronic communications faster, smoother, less frustrating, more effective. While information on a real-time basis is certainly a mission-critical need in today's business environment, what's overlooked in all the scramble for digitally based solutions is an even greater communications need: the communication between individuals.

Organizations that bring in Shared Values create shared standards of conduct and behavior around meeting behavior, proper use of e-mail, phone etiquette, customer contacts, interdepartmental or team communications, agenda development for meetings, and the myriad other forms of communications we engage in every day.

In the communications age, every person in the organization has to have the ability to communicate effectively. And, yes, even people who have never figured out where the commas belong and don't know that "between you and I" is a horrible mangling of English grammar can learn to convey ideas succinctly, logically, and convincingly. A "full-court press" on communication skills needs to become a part of the long-term development we urge on our employees.

Leadership and Citizenship

Tom Peters says in his excellent and commendable business video, *Leadership Alliance,* that he went out looking for four leaders, and found 4,000—by which he meant that in highly effective organizations, everyone takes responsibility, everyone makes a full-time contribution, and in other senses fulfills the roles that we associate with leadership.

Looked at another way, what we're talking about here is the issue raised earlier of citizenship versus leadership—that not everyone believes they can become a leader, but everyone knows they can be a contributing member of the society, and this truism applies to the society called the enterprise as well as to the social organization called the nation. Each demands the same basic aptitudes: courage, service to others, duty, selfless behavior, responsibility, teamwork, involvement, communicating openly, thinking things through, making decisions, applying one's talents, leading, supporting as a follower, the willingness to sacrifice and take personal risks.

Researchers at 3M show they recognize the value of acting as corporate citizens when they take the company up on its offer to let them spend as much as 15 percent of their time pursuing a pet project of their own. Acting as citizens, they are also showing the qualities of leaders.

As pointed out in the earlier discussion of leaders and followers, what we need today are organizational citizens. If you ask the average employee, "Can you become a leader?" the response will be immediate and without hesitation: "No." But a leader is someone who shows the way, and this can be understood in a much broader sense than just referring to those who dwell in the executive suite. Yet since people will persist in denying their potential to become leaders, we substitute instead the term "citizen"; the characteristics are largely the same: everyone aspires to citizenship and considers himself a citizen of the organization.

And leadership, it turns out, doesn't even require the "born to lead" qualities some would have us believe. Andrew Grove, CEO and a founder of Intel, condemns the misconception that effective leaders are superhuman and different from the everyday middle manager. Grove once told a *Fortune* magazine journalist, "Leaders are individuals who make ordinary people do extraordinary things," and cited the case of a subordinate who needed to develop leadership skills. He wouldn't get there by reading books about great leaders, Grove insisted, nor by going out on a wilderness program or taking a "leadership development" course. Instead, he would have to learn to be a leader "the same way each of us learned the important, unteachable roles in our lives—by studying the behavior of people who have made a success of it and modeling ourselves after them."

The chief distinction between citizenship and leadership lies in the aspect of an articulated vision. In their 1985 book, *Leaders*, Burt Nanus and Warren Bennis tallied the results of decades of study and research and came up with an astonishing 350 definitions of leadership. They found "no clear and unequivocal understanding" to distinguish leaders from nonleaders, effective leaders from ineffective leaders, or effective organizations from ineffective organizations. "Leadership," they wrote, "is like the Abominable Snowman, whose footprints are everywhere, but who is nowhere to be seen." Nonetheless, they offered this list of competencies that they believe embody leadership behavior:

1. Attention through vision: leaders create and unite others to their vision.
2. Meaning through communication: leaders are able to communicate their vision to others.
3. Trust through positioning: leaders create a confidence in the future of the organization.
4. Deployment of self through positive self-regard: leaders are aware of their strengths and weaknesses and the appropriateness of their talents.

An organization's philosophy regarding leadership and citizenship can be articulated and taught. The process begins with the selection in hiring, and should be made a part of orientation, and furthered by being built into the compensation system. For example, if long-term development has been adopted as a goal for all employees, this becomes part of the definition of good corporate citizenship, and can be rewarded in the pay program.

Ongoing Quality Initiatives and Measurement

Not only do we want to have long-term, universal learning incorporated into the work environment, but we want to take a look at our processes and procedures, and how they relate to building quality into everything we do.

One of the obvious ways to measure and quantify your work environment is to periodically survey both your people and your customers. Jack Welch of GE suggests there are three barometers: cash flow; your people's attitudes about their work environment and the company; and your customers' attitudes about your products and services. Measuring those three on a continual basis will have a great sobering effect on you and your business.

How do we create "leanness" in processes, yet with an abundance of redundancies in the way in which we serve customers? Think of Nordstrom, where the customer can return a pair of shoes or a bathrobe with no receipt in hand, and it's only after the customer leaves smiling that the employee deals with the details of the return.

The goal: an over-emphasis on service, and a very elegant delivery system to the customer.

WRAP UP

Your aim should be to integrate the eight People Systems into a cohesive, interrelated system, with all eight functioning smoothly together. Again, the eight People Systems establish the new Operating System context:

1. Hiring
2. New member orientation
3. Compensation, incentives, and motivation
4. Personal improvement, coaching/mentoring, and feedback
5. Long-term career development
6. Communications information transfer
7. Leadership and citizenship
8. Ongoing quality initiatives and measurement

One word of caution about the appropriate order of business: to avoid resistance and a lower level of response, it works much better to establish Responsibility-Taking, Decision-Making, Consensus Building, and the other techniques on the People Values side of the equation before starting to introduce the members of your organization to the concepts of the eight People Systems.

How do you actually go about kick-starting these eight People Systems developmental programs? That's the subject of the next chapter.

14

Rolling Out the People Systems Action Plan

You can dream, create, design, and build the most wonderful place in the world, but it requires people to make the dream a reality.

Walt Disney

This chapter presents an action plan that will primarily be of interest to the people in your company who are directly involved with putting the People Systems into effect. It's based on techniques we've evolved and refined over a period of years. Not the only possible approach, of course, but one that has demonstrated its value for companies small and large.

THE GROUNDWORK

Once the organization has decided it wants to integrate the eight People Systems, how do you roll them out?

The first step of launching a People Systems effort begins with identifying a "point person"—the project leader who will drive the entire effort. If not the CEO herself, it needs at least to be someone who is very committed to the effort, and who has a direct line to the CEO—the VP of operations, the head of

HR, a plant manager. Experience shows that the effort will be much more successful if the person leading it is someone who has high credibility among company employees.

As an initial step, the CEO (for convenience, we'll assume here that your CEO is the person who is driving the effort) will lay the groundwork for the process by providing a brief statement of the philosophical underpinnings of the organization. Not always an easy process, this requires thinking about the organization in terms quite different from the way we usually do. Many of us, when we go to church or synagogue, shift into a different plane of thinking than we use the rest of the week; in the same way, the CEO or other leader may—like most of us—be unfamiliar with thinking about the organization in this vein, and it sometimes proves desirable to form a small group of advisors who will be able to readily guide the CEO in formulating his or her thoughts—an important step since it sets the thinking for all that follows.

The paper that the CEO draws up will typically be just one or two pages, beginning with a preamble; a couple of examples will suggest how widely this preamble may vary from one organization to another:

> We believe people want to be great, to succeed and to grow. With this growth comes stretching and risk taking. We also believe the organization has a responsibility and a partnership role to play with the individual in his or her personal development.

Or:

> We believe people need to be flexible, capable of taking on new and important roles in the organization. Our business is dynamic, requiring people with high energy and a vision of the future.

The paper will then go on to set a course for the effort by providing a similar kind of statement about each of the eight People Systems. These need not be elaborate—a short paragraph on each will serve the purpose. For example:

Communications need to be simple, clear and direct; and in making decisions about communications, all those people who will be affected need to be included.

In other instances a longer explanation may be necessary.

At this meeting, the CEO will also select leaders who will assemble and guide task forces that address each of the eight People Systems; again, these are:

1. Hiring
2. New member orientation
3. Compensation, incentives, and motivation
4. Personal improvement, coaching/mentoring, and feedback
5. Long-term career development
6 Communications information transfer
7. Leadership and citizenship
8. Ongoing quality initiatives and measurement

All of this—reviewing the paper and naming a leader for each of the eight task forces—can usually be accomplished in a single meeting of no more than two hours.

Involvement of CEO or Other Project Leader

EVENT	ESTIMATED TIME
Planning meeting—define philosophy, set objectives	2–3 hrs.
Meet with task force leaders— create curricula	2–3 hrs.
Monthly progress reviews	45 min., with each task force leader
Three-month review	30–60 min., with all members of each task force
Final report	2 hrs., with all members of each task force

ORGANIZING THE TASK FORCES

Who will make good task force leaders for this kind of effort, which requires both understanding and a sensitivity to the needs of individuals? Based on years of guiding companies through this process, we recommend these yardsticks:

Task Force Leader Selection Criteria

- **A minimum of several years with the organization**

- **Credibility among other employees**

- **Good verbal and writing skills**

- **Good organizational skills**

- **Comfort with facilitating different viewpoints**

Note that while most people selected as task force leaders will be recruited from the ranks of managers, this is not a requirement, and frequently not the best approach.

Once the CEO has named the eight task force leaders, they meet with him in a session that introduces them to the road ahead and their role.

Ordinarily task force leaders will tap the people they want to serve as members of their group (eight to twelve is the recommended size), but a quite different approach has worked well at many companies: the organization throws some kind of everyone-invited event to announce the People Systems effort, describe the goals, and introduce the philosophy statements. Then those interested are invited to volunteer for membership on one of the task forces; and despite what you may have heard about people not volunteering, task force leaders usually find strong interest, and a choice of candidates.

For a multisite organization, will you have one set of People Systems, or will you need a different set at each company location? You may decide that allowances need to be made for the different cultures and relationships that exist among, say, offices in Georgia, North Dakota, and California; or between corporate and the regional sales offices; or between a *maquiladora* in Mexico and a highly automated factory in Salt Lake City. If you decide on multiple sets of task forces, the effort will inevitably demand a good deal of coordination among the different groups.

Whether the decision is to go with multiple sets or just a single set, experience suggests that the best solution for the multisite organization often lies in developing a core group of processes that are universal for the company, and then having additional location-specific items developed by the local teams as needed, to address the issues unique to their own region, site, or focus.

THE TASK FORCES AT WORK

Fight the temptation to dive right into the practical realities of moving the project forward. First things first: there needs to be a clear sense of where the task forces are trying to go. In our consulting practice, we recommend one-day kickoff sessions attended by all task force members—*sessions*, plural, because we find that there's better interaction if the maximum size of the group is limited to about 40 people; often this means holding two or more duplicate kickoff sessions.

The curriculum we recommend for these sessions covers these topics:

Philosophy of the People Systems
The eight People Systems
Overview of the rollout
Rollout schedule (developed by the group)
The task force approach and the five-step Decision-Making
    model (already familiar to employees from their
    participation in teams while learning the decision-making
    aspects of Shared Values)
Brief outline for planning the redesign (created by the
    group)

Following the kickoff meetings, the individual task forces are ready to begin their work. They will meet preferably once every week or ten days, over the next several months. The secret for a successful effort lies in making progress on each of the eight People Systems at the same time—with the exception, as you will recall, that *hiring* provides the driver for all the others, because the qualities we want in the employees we hire establish the criteria for the way the company wants to function. So the hiring task force needs to get something of a slight head start, and provide guidelines to the other groups as early as possible.

Though you'll be developing all eight systems in parallel, they will eventually need to be rolled out to employees one at a time—lest the effort of learning and adapting to the new ideas detract from the day-to-day needs of running the business. So early on in the process, priorities need to be assigned: in what order will the eight systems be introduced?

A related question requires you to think about the strategy for integrating the results of the eight task forces. This calls for some thought up front, even though the details won't become clear until quite late in the process—shortly before the results are reported to the project leader.

The main efforts of each task force revolve around the ten steps of planning the redesign:

1. Introduce goals of the task force; gain agreement from members.
2. Identify the information to be gathered, and gather it.
   Most often this will take the form of a small survey of employees, asking, for example, "In your view, what kinds of people should we be hiring?"
3. Establish philosophy based on the findings.
   "What is the philosophy of this organization regarding career development?"
4. Outline the proposed new process.

This will be a reflection of the stated philosophy of the organization, and what employees say they want, tempered by the realities of what is possible.

5. Design or map a process.

   The process map developed by each task force will be aimed at answering key questions such as "How are we going to develop citizenship?"

6. Construct a flowchart of your process.

7. Begin writing an overview statement of objectives.

   This statement, which will accompany the flowchart, describes what will be achieved in each step.

8. Gain agreement on the philosophy and confirm objectives; solidify the process schematic.

9. Complete the process design.

   Establish procedures, forms, and measurement tools. For example, the hiring team will create form letters for hiring and for rejection.

10. Prepare for presentation of results.

INTEGRATION AND REPORTING

While the eight groups work separately, they interact closely throughout the process: each provides notes monthly to all of the other groups outlining its progress. Each task force leader meets once a month with the CEO or whomever is guiding the overall project to report on progress and answer challenging questions; in many companies, the Human Resources director also attends these sessions.

It was emphasized at the outset of this chapter that when a People Systems effort lacks complete involvement from the top, the effort is likely to bog down under the demands for attention from more glamorous projects that hold the tempting promise of short-term profit. Despite the many distractions, your CEO or project leader needs to make time for these monthly meetings with his task force leaders—to demonstrate ongoing commitment, of course, but also to provide the kind of guidance and advice that his experience makes possible. Experience shows that these sessions are best scheduled for about 30 to 40 minutes each. Along with the verbal report, each leader presents a written update.

These monthly reports are typically held over a period of five months. In month three, we recommend that the regular monthly meetings with the CEO and HR director be replaced by sessions not just with each team leader, but with the leader and his or her entire team—so that all team members have the opportunity to share input from the CEO directly. We call this special meeting "the course correction."

FINAL REPORTS

At the end of the development period, the members of all eight groups will gather and share the results of their work—which should include distributing printed copies of the final summary report to each attendee; many companies also distribute this summary report to all employees. (We recommend that presentations be done, not by the group leaders, but by members— once again taking advantage of the opportunity to extend the experience and capability of individual employees, "growing" the individuals.)

Every gathering of any size is liable to have people who inadvertently offer comments, questions, or suggestions that seem to undermine the efforts of others. Here's one effective way to avoid this unsettling and destructive situation: instead of opening the floor for discussion after each presentation, arrange for all comments to be submitted in writing. Each group prepares a feedback sheet, which attendees use for submitting their comments. These feedback sheets should provide space for three types of input:

Comments on the basic proposal
Comments on the direction
Comments on particular elements

Also, encourage attendees to write their comments in the form of "more" statements and "less" statements—"More emphasis on hiring Native Americans"; "We should be encouraging people to take advantage of on-site higher education courses by giving less credit for courses taken off site."

THE STRATEGIC TASK FORCE

The presentation brings to an end the work of the eight task forces. The task forces are disbanded, and the task force leaders come together to form a new group, the Strategic Task Force, which is chaired by the CEO, an HR professional, or other project leader.

This group reviews each of the eight plans and the feedback comments submitted at the review session, and from all this assembles an action plan including completion dates. As its major task, however, it will develop a *People Systems Handbook* comprising the forms, policies, and procedures prepared by each of the task forces. In addition, it may also need to launch an effort to purchase or develop one or more curricula, if called for in the task force plans.

The Strategic Task Force then follows through with putting the People Systems into effect throughout the company, an effort that will typically take three or six months.

WRAP UP

Companies that have already launched a Shared Values environment offer a caution: from initial introduction to a fully functioning process throughout the company had taken twice the amount of work and personal time they had planned for or expected. And the stumbling block was those elements that this chapter deals with: the eight People Systems.

Why had People Systems turned out to be so much more of a problem than anticipated? Managers came to realize that their organization had ignored and shunned what should have been faced years earlier. Issues of behavior, accountability, responsibility, self-expression, and self-esteem. Issues of control, trust, truthfulness, empowerment, advancement, fairness. Barriers based on gender, race, age, culture. Uncertain standards, strategies, and priorities. Conflicts caused by different styles of managing, turf-building, politicking. . . . And the list goes on.

These issues, unrecognized, had been undermining the organization for years, until the Shared Values philosophy became the catalyst for facing and resolving them.

So be realistic in your planning: allow five or six months for the planning and another three to six months for finishing the *People Systems Handbook* and beginning to roll out the eight new approaches.

One final effort now remains for completing the Shared Values Operating System in your organization: redesigning the organization around serving the customer—in a seamless way.

Customer Processes

15

Creating a True
Customer-Centric
Organization

There is only one boss: the customer. And he can fire everybody in the company, from the chairman on down, simply by spending his money somewhere else.

Sam Walton, founder, Wal-Mart

FOCUSING ON THE CUSTOMER

The Shared Values Process Operating System culminates with this final model for organizational redesign. If you have been following our recommended timetable, your organization now has people who are practicing Shared Values—building consensus, experimenting with other people's ideas, taking responsibility in an appropriate way, making decisions based on gathered information, and so on. And once those are in place, the organization will have begun moving toward redefining its People Systems.

But—yes, there is a *but*—the efforts so far have been *inwardly* directed, focused on the organization itself, the people, the ways people relate to each other, and the ways the organization relates to people. What's missing? Of course: turning the focus outward to relate to any company's most important asset, its customers.

Reengineering also offers a focus on customers as a major goal. The difference between the two approaches, though, is significant.

The Shared Values redesign aims, not to reduce redundancies, but to get employees closer to customers' needs. The difference can be illustrated by a real-life incident: several years ago, a couple walked into a Sears store and approached a clerk who was lingering between the televisions and the kitchen appliances. The man said, "We're here to buy a washing machine and a TV." To which the clerk replied, "I can sell you a TV, but you'll have to find someone else to sell you the washing machine."

In a reengineering or with TQM, there would be no change in the store clerk's authority—it would be determined that customer needs were being met, and the goal would be to streamline both departments; no provision would be made for customers interested in "shopping the store"—wandering through looking for things that catch their eye or buying things from two or more departments. The result would be an arrangement not designed for customers, but for the convenience of the organization.

A document has circulated in the business community that purports to be an actual policy statement approved by the Board of Directors of General Motors in 1972, at a time when the small Japanese cars had not yet established more than a toehold in the U.S. market. Whether a real document or a much later attempt to poke fun at GM's lack of foresight, it has become an accepted piece of business dogma, an amusing reflection on a failure to understand one's customers. In part the statement reads:

1. GM is in the business of making money, not cars.
2. Success for GM will come from having the resources to quickly respond to innovations introduced by others.
3. Cars are primarily status symbols. Styling is therefore more important than quality to buyers who are, after all, going to trade up every year.
4. The U.S. car market is isolated from the rest of the world. Foreign competitors will never gain more than 15% of the domestic market.
5. Energy will always be cheap and abundant.
6. Workers do not have an impact on production or product quality.
7. The consumer movement does not represent the concerns of a significant portion of the U.S. public.

REDESIGNING THE ORGANIZATION

The redesign issue can be illustrated with an example: a customer in Manhattan brings a package into UPS and asks for same-day delivery to Los Angeles. How does UPS respond? At most, the clerk might point the customer toward one of the airline package services, but that would be the extent of the help offered. Any attempt to handle the package would disrupt its processes, and in the end "underserve" the customer when the package didn't arrive on his timetable. Defining limits is just as important as creating seamless customer service.

What Shared Values calls for, though, is no small challenge. If you were to begin redesigning UPS or a chain of Sears-like stores from the customers' perspective, you would give your process engineers apoplexy. The reason: truly serving the customers leads to building in a lot of redundancies and lots of provisions for handling special situations. It also leads to dispensing with entire shelves of rules and procedures, instead turning a lot of decision-making over to a reliance on the employees' good sense. But good judgment alone isn't enough; employees would also have to understand the cost of goods so they wouldn't give away the store. This wouldn't be just a re-design, it would be a *revolution.*

WHO IS YOUR CUSTOMER?

For a company in manufacturing, getting closer to customers would almost certainly require the at first glance drastic contra-dictory act of reducing the number of customers. In fact, most manufacturing organizations that have implemented Shared Values have realized that 40 percent or so of their business was with customers they should not have been doing business with. Why? Because what they were selling those customers was beyond the firm's parameters; for example, requiring the manu-facture of items that did not spring from their core competen-cies. On examination it was clear that, no matter how hard the organization tried, it was underserving the customers. But the worst part was that these customers were unprofitable to the company.

Once a cross-functional task force has much more tightly defined who the customers are that the company should really be serving, and has pared down the customer list, then the or-ganization is able to focus on serving the remaining customers at a much higher, more immediate, more personal level, not to mention more profitably. Organizations that are contemplating ISO 9000 (or, for the auto industry, QS 9000) will be helped greatly by first establishing a customer-centered organization, before they go down the road of cataloging their processes and procedures.

EVOLVING THE PHILOSOPHY FOR BECOMING CUSTOMER-CENTRIC

The approach here parallels the approach to the People Sys-tems: before tackling the practical aspects, begin by analyzing and crafting the philosophy that defines the underpinnings. If your company is to become truly customer-centric, what will your organization have to look like? Our experience after years of helping companies with their redesign efforts convinces us that those companies which address the tough customer issues are the ones able in time to differentiate themselves from their competition.

These decisions need to be tempered in the furnace of reality. It's easy to announce that your company will be committed to "providing uncommon service" to customers. But how will this be translated into practice, and is your organization truly willing to spend the additional amounts it will cost, say, for guaranteeing same-week shipment of orders; for guaranteeing Saturday and Sunday delivery; for providing 24-hour, 7-day customer support?

The decisions for each business, regardless of its industry, are driven by the marketplace. When the competition raises the bar, will you be able to rise to the challenge? Will you be able to raise the bar first, making others rush to keep up? How will the bank be able to stay open until 9 every evening? How will the trucking service manage to offer same-day delivery, or reconfigure its facilities so it can handle a customer's shipments regardless of size? How will the chain of discount stores be able to provide enough staff training so that clerks are well equipped to handle all shopper questions, storewide? How will the machine-tool manufacturer afford to offer a trade-in policy for upgrades?

In the most successful efforts, corporate leadership works in concert with a cross-section of the organization's people to identify the parameters of the philosophy. Once the leadership has hammered out a version it can subscribe to, it seeks input from the members of the organization. The philosophy statement it has developed goes to the group that's going to be directing the efforts of creating the Customer-Process Action Plan—the plan which will translate the philosophy into practical terms. As in the development of the People Systems, this statement of philosophy must not be thought of as carved in stone at this point: it must be accompanied by an invitation to challenge every aspect—each assumption and each statement. This process has become known in management-theory circles as a *semi-consensus-building approach.*

By the time the organization arrives at the end of the process described in the next chapter, all employees should have had the opportunity to challenge, study, internalize, and ultimately support the philosophy—not because they are told to but rather because they see it as a desirable thing to do.

Speed is of no importance in this process. The ultimate value lies in the buy-in by all employees.

Note that high-risk strategies are a danger unless the people of the organization truly resonate with them. The best statement of philosophy is simple, straightforward, and establishes a strong context. Allow for broad interpretations, entrepreneurial spirit, and an abiding enthusiasm for what the people of the organization are doing.

It should also be your goal to create a statement of philosophy that causes every employee to sense a personal fiduciary responsibility.

A great statement of philosophy creates flexibility, not rigidity.

TOWARD CREATING THE CUSTOMER-CENTRIC COMPANY

Once the organization has defined a customer philosophy that is both realistic and achievable, it's ready to address the practical issues of redesigning processes that will bring competitive advantage. When an organization figures out how to do this, it positions itself to gain the preeminent position in its marketplace. Whether it's Snap-On Tools, Southwest Airlines, Price-Costco, or Starbucks Coffee—an organization that becomes customer-centric establishes a competitive edge.

A caveat: when an organization moves from processes that support customers to processes that are *efficient,* customers will walk away. For a prime example of this truism, recall what happened to General Motors in the 1980s. The company adopted a policy of designing its cars around interchangeable engines and components that were identical across several different nameplates. More efficient, sure—but the vehicles no longer seemed sufficiently unique to car buyers.

Lessons like that weren't lost on the founders of Eagle Hardware. Thinking about what they wanted their business to be, they studied the marketplace and discovered a prime piece of information: people would go out of their way, driving an extra distance, to shop at a hardware store where the clerks were knowledgeable enough to give advice and recommendations to help the customers shop. Price and the variety of product on the shelves were important but not enough; it was too easy for the competition to steal their thunder. Armed with this insight, they designed a customer-friendly environment with a knowledgeable, well-trained staff. (One industry study reported that over a lifetime a home handyman will spend $47,000 at the hardware store—a great blessing to the store that can secure that shopper as a steady customer.)

Starbucks Coffee saw a decline in coffee drinking among the demographic group targeted by Maxwell House and other coffee leaders. But tiny Starbucks went after this group anyway, convinced by research that these consumers weren't being offered what they would find most appealing. Starbucks offered high quality and a variety of choices, and created a context that positioned the coffee experience as a social event. And it worked: with more than 1,000 outlets, Starbucks has become the largest U.S. retailer of gourmet coffee. Its stock outperforms the market by a wide margin—in the 12 months between

February 1996 and the time of this writing, Starbucks' shares increased in value from 18 to 32, along the way hitting a high of 40 and looking more like a software company than a coffee chain.

THE ELEMENTS OF CONTEXTUAL CHANGE

The steps for creating a Customer-Process Action Plan, as detailed in the next chapter, focus on two basic mechanisms that are used to create the contextual change you are seeking: a "charter" and an organizational design.

The charter serves the same function as the constitution of a nation in providing a framework for the fundamental beliefs of the people.

Along with the charter, you will also create a design centered around how customers can best be served. The goal here is to define how the organization needs to appear and respond *when viewed from the customer's perspective.* If you are an auto dealer, you have New Car Sales, Used Car Sales, Service, Parts, and Finance. These are the departments; a customer may arrive with a price in mind, not sure if she'll be better served by a new car or a used one. The next customer through the door may be looking for advice to help him decide whether to buy a part and install it himself, or leave the car for service. At a traditional dealership, both these people will run up against compartmentalization. Can you redesign the organization so it better serves the customers' needs?

Large corporations suffer from this problem in the extreme. Tom Zusi, VP of Finance and CFO for a division of AlliedSignal, describes a situation that will sound familiar to many other companies: "We used to do business with one airline that had to deal with 17 different parts of AlliedSignal. They got 17 different invoices, which they paid to 17 different places. That's not very customer-friendly." And he adds, "Our strategy today is to be transparent to the customer and to be one company. We've got to be seamless to the customer." That's a great statement of the goal in the design effort of making your organization customer-centric.

In Chapter 16, we look at the step-by-step techniques of rolling out the Customer-Process Action Plan.

WRAP UP

It's become trite to say "Know your customer." But that's where the process of evolving a Shared Values organization into a *customer-centric* Shared Values organization needs to begin.

It makes no difference how long it takes to achieve this final phase of creating a true Shared Values environment. Speed isn't the issue; rather, the issue is achieving an organization that wins, and retains, the support of its employees and its customers.

This last phase begins with creating a statement of philosophy for the customer-centric organization you are aiming to achieve. A significant and workable statement of philosophy creates flexibility, not rigidity. When this is accomplished, policies are limited, possibilities are limitless, and lasting change becomes a reality.

16

Refocusing the Organization into Customer-Centric Process Groups

In every instance, we found that the best-run companies stay as close to their customers as humanly possible.

Peter Drucker

Every crowd has a silver lining.

P. T. Barnum, showman and circus owner

THE GROUNDWORK

As with the detailed instructions for establishing a People Systems Action Plan (Chapter 14), the procedures described in this chapter will primarily be of interest to those in your company who are directly involved with designing the Customer Process and putting it into action. Again, while obviously not the only way to proceed, the methods described in this chapter are based on techniques that our specialized consulting group has evolved and refined over a period of years.

Companies today talk about becoming customer focused, customer-centric. It's a goal of virtually every up-to-date business person who is motivated by competition. Fine—but how do you actually go about achieving this? The answer we offer lies in refocusing the entire organization into "process groups" that take a customer-centered approach; this involves not just those people who have direct contact with customers, but virtually *everyone* in the organization. (See illustration on page 234.)

As a first step toward achieving this new kind of customer-focused structure, the CEO and leadership will develop and publish a White Paper outlining the philosophy and the parameters for the redesign initiative. Typically this document will be about three to five pages and will begin with a vision statement expressing how the organization intends to serve the customer. The paper then defines what business the company is in. (This sounds obvious; a famous example points out that if railroad magnates earlier in this century had perceived they were not in the railroad business but rather in the business of moving people and goods, they would have considered expanding into all general transportation categories and might now dominate the airline industry, shipping, and package express delivery instead of being left behind with only a small, withered part of a huge industry. For more on this, see "Marketing Myopia," by Theodore Levitt, *Harvard Business Review,* September/October 1975.)

The White Paper will then examine future opportunities in the marketplace, and how the organization intends to differentiate itself from the competition.

Process Group Redesign
Focus On Each Customer

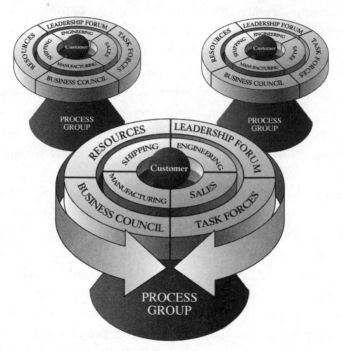

Each customer has a group to support his or her product and service needs. The number of process groups depends on the organization's size.

ORGANIZING THE STEERING COMMITTEE

The demanding but rewarding challenge of developing the Action Plan will be carried out by a Steering Committee—a cross-functional group selected by senior management. Although its work will require meetings over a period of about four months, everything that follows rests on the quality of its effort. Moral: pick good people.

As we've already seen, the committee will write a charter and craft a new organizational structure, and lay the plans for a restructuring of the organization centered around customer processes.

The Steering Committee, as with the People Systems effort, is chaired by the CEO or by another senior executive such as a general manager or plant manager. For the multisite organization, you will again need to decide whether to use a global committee serving the requirements of the entire company, or several committees, each serving the narrower needs of separate parts of the organization. Many organizations find that elements like payroll and certain purchasing functions can be handled on a global basis, while other elements need to be locally determined.

Our experience suggests that the most effective committee size is 18 to 25 people; in fact, in organization after organization, we have found that 22 people seems to be a magic number. This size simply works better than smaller or larger; the reason isn't entirely clear, but it's our empirical truth. (Although the committee initially splits to work as two separate units, the "magic number" has its impact later when the committee reforms to work as a whole.)

As for membership in the Steering Committee, our recommendation is likely to come as a surprise: the group works best when about one-third are senior and middle managers, and the rest are *non-management* employees; the reasons should become clear from what follows.

The Steering Committee at its first meeting divides itself into two groups—one that will work on the charter, another that will work on the design. Both are driven by the ideas expressed in the White Paper.

THE DESIGN GROUP AND THE PROCESS TEAMS

The design group wrestles with many details of the organization, all focused around one central concept: reorganizing the work effort so that virtually all tasks and projects—especially those that directly impact the customer—are handled, not by individuals or departments, but by "process teams."

In 1985, a 747 jumbo-jet crashed in the mountains outside Tokyo's Narita Airport, the worst crash in the history of aviation. After extensive probing, the investigators found that the pilots had been highly skilled and had apparently made no flight errors, the weather had been clear, the 747 had never had a similar accident (nor has it since), and no errors had been made by air traffic control.

Aircraft maintenance had been performed as required, yet a section of the rear fuselage, recently repaired, had ruptured, knocking out the elevators and vertical tail, and with it the ability of the pilots to control the aircraft. In the end it seemed clear that each individual with a maintenance role had done his or her job, but that no one had been assigned maintenance responsibility for the aircraft.

The policies are changed now: each aircraft has a maintenance group assigned to it, and each person working on an aircraft knows that he shares a personal responsibility for it. At the end of an overhaul, the team shows its confidence and commitment to the work it's done by sending one of the members on the plane's next flight.

The team concept has kept the airline free of any repetition of the Narita tragedy. Not only that, but the new approach has even benefited the bottom line: the airline has been able to reduce maintenance costs by over 15 percent annually.

One goal of a process-team approach is to make certain that customers with problems can get the ear of an individual, or when necessary an entire team; routine matters are handled smoothly, and nightmare problems are dealt with responsibly.

Your company can look forward to knowing that once this structure has been installed and is running effectively in your organization, every company or individual you retain as a customer will have its own process team; every customer call, need, question, and problem will be handled by someone who knows the customer, knows what products they use, their special requirements, ongoing issues, and their history with your organization. And on your side of the equation, each employee has a full support team as a resource.

The challenge for members of the design group, easy to state but not so easy to achieve, requires that they determine what will make your business easy for the customer to deal with. For models, look at the philosophy that drives British Airways, and Southwest Airlines, and Starbucks Coffee; then look at how each conducts its business. For examples of what to avoid—well, we all have our own favorite candidates.

THE CHARTER GROUP AT WORK

When completed, the charter will describe a new management structure for the organization and the mechanics of how the organization will behave—all focused on creating a company that will truly be able to serve the customer. As a major part of its work, the charter group also defines the parameters for a new "Business Council," which will become responsible for the organizational goals and mechanics prescribed in the charter.

The Business Council will be headed by the CEO, but its members will be *elected* by all employees.

In the past, the tasks of the charter group and the Business Council would have been responsibilities of management. This represents a major element in the shift by the organization from a hierarchical to a *representational* structure.

If you are not yet convinced that this could work, consider the example of the Saturn division of GM—where it uses an organizational structure in which the union workers, central management, and independent Saturn dealerships share responsibility for making decisions that impact the entire organization; the success of the Saturn operation, and the delight of Saturn customers, is already looked on as one of the bright examples of enlightened management, and will no doubt be the subject of business-school case studies well into the twenty-first century.

If you're still not convinced, go back and reread the story of Johnsonville Sausage, in Chapters 9 and 10.

The approach succeeds because it relies on the wisdom of people who are committed and are well educated about the business they are in.

Over the years that we have been helping companies install a Shared Values environment, we've seen that employees are clear about what they know and what they don't; and when they don't, they ask their peers, coaches, and Wise Counsels, the CEO, plant manager, the company's accounting firm, outside consultants, or whoever else can help them arrive at the best solutions.

Employees take this responsibility very seriously, and it's heartening to see how excited many become about being accepted in a way that acknowledges their greater value to the business. True story, from one of our client companies: a truck driver went home and told his wife, "I'm now on a task force to select a new inventory and accounting system for the company." "But you didn't finish high school," his wife pointed out. "You don't even balance our checkbook, and they're going to let you have a hand in something like that?" When this man tells the story, it's clear he will never again see himself as just a truck driver.

The roles of the Business Council, which need to be spelled out in the charter, are these:

1. Establish and protect the organization's values and behaviors—such as product quality, service levels, ethics, and citizenship.
2. Establish budgets, monitor and report financial achievements, and take reporting responsibility for budgets, expenditures, and cash flow from the process groups.

Typical of the change that will take place: in the past, when a new copying machine was needed, purchasing would get bids, evaluate them for best price, performance, quality, and service, and place an order. In the Shared Values environment, with a functioning Business Council, the people who will use the copying machine analyze their needs and requirements, and—within budget boundaries—make the buying decision on their own. Organizations come to recognize that users need to be consulted; the process described here lets the people take ownership—driving decision-making down into the organization.

3. Promote process groups and make sure they work.

4. Champion continuous process improvement in quality, customer Shared Values, and productivity. Quantify improvements within the process groups.

5. Monitor, support, and carry out the People Values and Business Values of the organization.

6. Be responsible for coordination and support of all resources to the process groups, making sure they get what they need.

This includes contracts, budgets, and financial goals, as well as making sure the groups communicate with one another, rather than functioning (or *competing*) as stand-alone units.

PROCESS GROUPS AT WORK

A process group will be formed for each significant process of the organization. Thus, in a factory, several machinery groups might be needed—each responsible for setup, maintenance, and repair of a number of machines, but, beyond that, probably also responsible for decisions about when to upgrade or replace the machines they're responsible for. Each of these teams might have subgroups for each of the three shifts.

In most cases, a process group or team will have cross-functional membership. How many teams are needed? Obviously a lot of factors impact the answer, but, typically, a factory of, say, 700 people might have 12 to 15 process teams.

The size of each team varies widely, as well. For tasks that involve direct dealings with customers, a single group may have only a few members and serve 30 customers, or may have many members and serve 3,000. The important issue is that customers always deal with someone on their own process team, and customers know this will be true every time.

Process groups provide an answer to the question "Where do managers go in a customer-centric organization?" Managers either become part of a process team, or become a resource or Wise Counsel to many of the teams. This is not a strategy to get rid of middle managers; rather, a customer-centric Shared Values organization redefines the role of managers—reducing their supervisory functions and converting them into Wise Counsels to the process teams.

WRAP UP

Shared Values is unlike any other approach—unlike TQM, unlike reengineering, unlike typical Human Resource solutions that aim to change people. The Shared Values approach teaches people how to improve the context of their work environment, leading to inevitable change in the processes and in the roles and responsibilities of all who work for the organization.

Most programs try to change people. This approach instead teaches people to change the philosophical context of their work environment, leading to inevitable change in the processes of the organization, and in the roles and responsibilities of all who work for it. And when that happens, the wings and tail section of the organization stop falling off.

The eight Shared Values offer a universal, unifying force

that speaks to all people in all cultures. For an organization that is established, mature, and functioning well, Shared Values helps clarify working relationships at all levels, pulling people closer together, reducing politics and wasted acts of turf-building. Shared Values helps create an organization without boundaries, where experience and tradition can finally begin mixing safely with new ideas; a new, improved attitude toward people; and a workable design for becoming responsive to people.

Organizations under stress, experiencing downsizing, market reversals, or financial pressures, will discover how Shared Values brings a dramatic healing process for lasting change.

THE PATH TO
LASTING CHANGE

17

Benchmarking Your
Work Environment

In advertising there's a saying that if you can keep your head
while all those around you are losing theirs,
then you just don't understand the problem.

Hugh M. Broille, Jr., NBC television

MEASURING

One of the largest ad agencies on the West Coast had been
plagued by problems in morale and in the way people related
to one another. The president had been struggling to improve
the situation by dressing up the benefits package, reasoning
that "These people are really driven by money." Finally he
threw up his hands; he had done everything he could think of,
but nothing had helped. "I'm all tapped out at this point, with
no act to follow."

When he decided to benchmark his organization with the
Values & Attitude Study, the results were, he said, like "a giant
can opener" for what they revealed. Quite the opposite of what
he had always believed, his people expressed concerns less

about money than about being respected, being part of the decisions, receiving more mentoring. They wanted more honesty, more consistency and believability from their leaders and managers, and more ethical conduct in everything from the numbers they put on their time sheets to the ad copy they were turning out.

The follow-up Values & Attitude Study, sixteen months after the Shared Values philosophy had been introduced, showed net earnings improved by 30 percent, while the Value Tension Index had improved by 29 percent, revealing that the company was indeed doing a better job of delivering what the employees wanted in order to feel fulfilled.

Corporate managers are prudently reluctant to adopt any program, process, or system that can't be measured. No matter how promising an idea seems, no matter how enthusiastically it's endorsed by people one respects, no reasonable manager would foist a new approach on her organization just on the basis of a personal reaction.

Yet measuring—benchmarking against some standard— proves to be the most difficult aspect of monitoring transformation.

That predicament is nothing new. It's always been very difficult to monitor and benchmark behavioral change. The quality movement, for example, incorporated strong measurement criteria for processes, machines, and output, but shied away from attempts to benchmark people. In fact, the entire human potential movement fails to address the issue of benchmarking, leaving its claims of success unproven and unprovable.

It's possible to put someone through a typing course and then monitor his speed and the number of errors. The same holds true with any mechanical skill. But measuring the value of behavioral or contextual changes in an organization is difficult because the variable is people. As a result, benchmarking organizational behavior change continues to be treated more as art than science.

Over the past 50 years, industrial psychologists have been trying to develop a suitable predictive device, and have created tools such as the well-known Myers Briggs Personality Profile. From the early 1950s through the 1960s, many companies asked psychologists to create tools to help them select people from a profile, hire them from a profile, and improve them from a profile. More recently, HR departments have turned to the so-called "360-degree" devices that evaluate an employee by asking those around him or her to quantify the person's behavior. But no devices can really predict future performance, and they're especially inept at accounting for variables such as personal life changes, change in career interest, a new work environment, a change of boss.

But what about the many custom survey tools being marketed to corporations? We believe they don't provide useful, insightful results, and yield no reliable benchmarking data, for three main reasons. Survey organizations frequently overlook questions on subjects of universal interest such as morals, racial issues, or pay issues. Most customized research tools, as well as standardized research tools used internally by HR departments, have no reference point—nothing against which to benchmark. And most of these research tools are not attached to a systematic approach that provides a systematic means of addressing each of the issues that the device uncovers. Unfortunately, most custom survey tools lead down an empty alley on the way to nowhere.

A major university medical school contracted for a customized employee survey; when the results came back, the president threw the summary on his top administrator's desk and demanded, "Get this stuff fixed."

But asking questions that are attached to a floating buoy in the middle of the ocean is unfair to everyone: the customized survey was not attached to norms; that is, the questions were not normalized against a broad base of data from extensive previous surveys.

> Worse, the institution had no clue how to act upon the survey results nor did it know what to do about the responses after it received them. The work had been done by a testing organization that only created survey devices and had no responsibility for the "fix."
>
> One year later, the report was forgotten and it was back to business as usual.

It's unfortunate but true that Mississippi high school students regularly score near the bottom on national achievement tests. If you were the principal of a Mississippi high school that ranked first in the state, you'd have reason to be proud of your teachers and students, until you stopped to consider that your school's rating against the rest of the nation was still embarrassingly low. Being first of the worst is not much to be proud of. (Like the old Henny Youngman joke: "How's your wife?" "Compared to what?")

In the corporate world, dysfunctional work environments too often go unrecognized because no one has effectively measured the correlation between their environment and the environment of comparable institutions. Every organization interested in performance improvements needs a measure that will clearly reveal these correlations: a company monitoring its own performance learns little; the results need to be seen in a larger context.

BENCHMARKING SHARED VALUES

One important challenge when building a lasting Shared Values environment is to apply a benchmarking tool and an indexing device that reflect and measure organizational behaviors against worldwide norms. This is especially important if your marketplace is open to worldwide competition.

Based on the 17 million survey responses that were the foundation for launching Shared Values, our team structured a benchmarking device, the Values & Attitude Study (VAS) described in Chapter 1. This is a 61-question, six-part survey tool. The VAS enables us to benchmark an entire work environment, and then compare the organization's environment against other company scores that have been gathered and recorded over the years and stored in our universal database.

Before an organization embarks on the process of installing a Values-Based Operating System, or for any organization not convinced that it needs one, an organizationwide study is highly revealing. For this, each employee completes the 20-minute pencil-and-paper VAS questionnaire.

The questions explore each employee's perceptions about the values that guide the actions, decisions, and behavior of people in the organization, in six categories: values, job satisfaction, people behaviors, styles, performance of the organization, and attitudes toward the work environment.

THE VALUE TENSION INDEX

Perhaps the most unique insight provided by the study comes from the triple response that the employees give to each question in the Values section: not just how important a particular value is to them, but how important they believe it is to the people who manage the company, together with their evaluation of how well the company is succeeding in this area. As explained earlier, the spread between how employees rate individual value-needs in the workplace and how well they perceive their organization is meeting those needs provides the rating we call the Value Tension Index (VTI).

A company that's doing an admirable job of creating an effective workplace finds only a narrow gap between what the employees say they need and what they say they are experiencing. A wide gap is a blinking red light alerting management that workers sense a severe lack in their work environment, the kind of lack that undermines success.

So the Index clearly identifies a company's level of need for inaugurating a values-based environment. More than that, it also provides a statistical jumping-off point against which to measure progress as the impact of the Values-Based Operating System starts to be felt. The Value Tension Index shows how good a job the organization is doing over time to narrow the gap between what people want and what they perceive they are receiving. Growing out of the premise that people can't be fixed or changed, the VTI measure shows, instead, how the organization is delivering to people what they perceive they need to be successful. And repeating the study every 14 to 18 months provides an ongoing gauge to show that the gains made through Shared Values are not eroding.

It's fascinating to note that a change greater than 3 percent has never been registered in the Personal Values Needs section of the VAS, but many improvements of 100 percent and greater have been recorded in the way respondents view their managers and their organization. A slight change in Personal Values leverages a significant gain in the other two categories.

Another aspect of the VTI as a benchmarking index can be illustrated by a story: two men each invite a friend from work to their homes for dinner. In Joe's apartment, Paul can hardly find a place to sit because yesterday's clothing and the kid's toys have been left all over the furniture. When Joe's wife finally calls them to dinner, much later than expected, the food is not just burnt but also, somehow, cold, and Paul can hardly wait to escape. The next morning at work, Joe is full of enthusiasm. "My wife isn't much of a housekeeper and she's a worse cook, but isn't she a great person!"

The following week Paul has dinner with his friend Tom and is struck by the contrast. The house is immaculate, the dinner of veal scaloppine is on time, and excellent, and accompanied by a good, dry Italian wine. Tom's wife, who has a job of her own, manages to be a gracious hostess and lively conversationalist at the same time she's getting a fine meal prepared and served without a fuss.

The next day Tom apologizes because his wife likes to cook fancy European dishes instead of nice, plain American food.

Joe thinks his wife, the meal burner, is world-class while Tom's wife, despite the candles and sauces, isn't fundamentally appreciated. Clearly it's Joe who would score better on the VTI rating—because what he wants and what he perceives as what he gets—the reality of his situation—are so closely aligned.

"HEIGHT" MEASURES AND WORLD-CLASS SCORES

Contrast the way employees of two different companies respond to the VAS questions about truth (again, ratings are on a scale of 1, worst, to 10, best):

COMPANY	HOW IMPORTANT IS TRUTH TO YOU?	WHAT LEVEL OF TRUTH DO YOU RECEIVE FROM THE ORGANIZATION?	VTI RATING ("GAP")
A	5.0	6.1	1.1
B	9.5	7.2	2.3

Which of these companies is in better shape? Recall that a VTI rating is, like a golf score, better when lower—a low gap indicating a closer alignment between the importance that employees assign to a value, and the extent to which they perceive their organization is delivering that value to them. Company A, with the lower score, looks better.

But in addition to looking at the gap, we also look at the "heights"—the absolute values, which in this case tell us that company B is, in the view of its employees, significantly better at delivering the truth. And, in fact, if company A has so little respect for truth as its employees believe, it's likely headed for the graveyard.

The reports we produce for client companies also show what we call the "World-Class Scores." These are scores that would need to be achieved in order to rank with the top 7 percent of all companies in our database for that particular value, behavior, or performance level.

OTHER MEASURES OF THE VTI

In addition to measuring the effectiveness of the organizational environment, the VAS survey also provides enlightening measures on a number of other aspects of employee needs and attitudes that spotlight problem areas in the workplace.

Empathy Score

Empathy involves identification with someone's situation, motives, anger, fear, or frustration. Empathy scores provide a measure for leadership in terms of how well organizational interests are being balanced against people's interests.

Managers have three choices every time they interface with an employee. For example, an employee goes to her manager and says, "I got this memo from headquarters; what does it mean?" The manager answers, "There they go again. I don't know what the home office means. I'll get back to you." A second employee asks the same question of manager number two, who replies, "We don't have to understand it, just get it done. You're already behind." But manager number three replies, "I'm not really sure, but next week I'll be at the home office; while I'm there, I'll find out and get back to you."

The first manager sides with the employee at the expense of headquarters. The second has no interest in seeking out a common ground between the employee's interests and the needs of the operation. Manager number three has the empathy to hear beyond the employee's expression of fact and pick up on the vibes of frustration, and shows an ability to balance the individual's needs with the organization's interests. This manager shows the kind of listening that it takes to become a true leader.

In the Values & Attitude Study, empathy is measured by

comparing how closely two factors align for the eight People Values:

The variance between the average rating by managers and the rating by all respondents

The variance between the average rating by managers and the rating of the organization.

If the variance is closer between managers and the respondents, this indicates that the managers are aligning with the employees, showing sympathies that are misplaced and inappropriate. If the variance between managers and the organization's delivery is closer, it shows the managers are aligning and sympathizing more, with the interests of the organization—indicating that they are not well in tune with their employees.

If the two variances are closely in line, the managers are able to empathize with both the people and the organization's goals and needs. This is the desired condition: it shows the balance that makes managers into leaders.

Levels of Satisfaction

Employees respond to a series of statements on their personal feelings about their work; for example, "I have personal control over my job performance"; "I am treated fairly"; "I am trusted by the management." They rank each statement from "strongly disagree" to "strongly agree."

When organizationwide responses are averaged and compared against worldwide norms, the company gains an understanding of how the employees view their work situation on a job-satisfaction level.

Levels of Performance

The measures for the level of performance gather employee opinions about how well the company is doing on a range of items including product quality, leadership, hiring the right people, employee reviews, and employee communications. This group of items offers insight into whether people feel sufficiently supported in these key areas.

Personal Style

Everybody in an organization functions in one of four basic roles:

ROLE	CHARACTERIZED BY
Hero	Selfless behavior
Maverick	High energy, problem solver, creative
9-to-5er	Comes and leaves on time, doesn't make waves
Dissident	Finds fault; nothing is ever quite right

In this section of the survey, employees are asked to decide which of these categories they belong in, and then decide what percentage of their co-workers belong in each of the categories. (The survey doesn't suggest any favorable/unfavorable character to the four roles; only in the report of results are the roles labeled.)

Most people see themselves as Hero or Maverick. By itself, that's a good sign. In an effective organization, people also rate *others* the same way—but that's not what most companies find. They find instead that employees rate themselves in one of the "good" groups, but think that half of the others in the organization are 9-to-5ers or Dissidents. The dramatic contrast between self-perception of employees and how the employees view those around them reveals an organizational context that is out of kilter.

In a healthy environment, the organization helps people self-actualize. If we hire a person who sees herself as a Maverick, someone who likes to fix things and solve problems with high energy, shouldn't she be allowed to behave that way in the work environment? Our findings reveal that 80 to 90 percent of the people companies hire describe themselves as falling into the categories of Heroes and Mavericks. But once they enter the work environment, the number in these categories dwindles rapidly to under 50 percent.

Why does this change in behavior take place? Because of the organization's context, of course. New employees quickly discover that being a Hero or a Maverick isn't safe, it isn't admired, and if they want to get ahead or just keep their job, they'd better change and adapt. If they don't figure it out for themselves, HR will soon tell them what course to take. So these potential problem solvers and hard-driving individuals become 9-to-5ers and Dissidents. Not a desirable trade-off, certainly, but a reality within organizations worldwide.

Job Attributes

In this final category of the survey, employees are asked to rank ten attributes in order of importance—including wages, job security, opportunities for promotion, interesting work, and manager loyalty. The top three attributes guide company leadership in focusing on key areas, or "hot buttons," that in the employees' view most need attention.

HOW BENCHMARKING SCORES FLUCTUATE

If you sign up for regular golf lessons, you hope that your game will begin to show a steady, if gradual, improvement. Your hopes are too often dashed. The fact is, of course, that even the tournament pros have some good days and some bad, some good tournaments and some bad. It would be reasonable to expect the same with Shared Values. As it turns out, though, a different dynamic is at work: during the third year after launching a Shared Values workplace, many of the organization's VTI scores may get worse, that is, increase. The second year improved but the third year did not.

When this phenomenon was first observed in the early days of developing the Shared Values Process, it raised concern that the approach might be running out of steam by the third year, suggesting that lasting change was, after all, not achievable. Research revealed that the cause lay elsewhere: in company after company, employees were seeing marked improvements, but by the third year were expecting their organization to reach for and achieve higher standards. Beyond the third year, however, results typically once again catch up with expectations. (See the chart on page 255.)

SELECTING A VENDOR

If you invite in any group to introduce an intervention such as an ISO 9000 program, transformation initiative, TQM approach, open book program, restructuring process, or reengineering approach, make sure it intends to benchmark, and will use a follow-up indexing device that is normed against other sites. At a minimum, the consulting firm should have a database containing study results on at least 100 other companies it has worked with in the past.

Talk to the previous clients. A key question: what action did these clients see as the result of their work with the consulting firm, and what benefits has it brought?

Before allowing any of your employees to take part in a sample study with a firm you are considering, evaluate the questionnaire it plans to use; avoid any firm that wants to ask blame-placing questions, such as leading employees to hold their manager responsible for problems in the workplace. (Sometimes the problem is indeed with the manager; usually it's with the processes, growing out of the vital role that context plays.) Our experience has convinced us that some 360-degree devices, IQ Tests, personality profiles, and the like have no better track record than newspaper horoscopes, because they attempt to predict how people will perform.

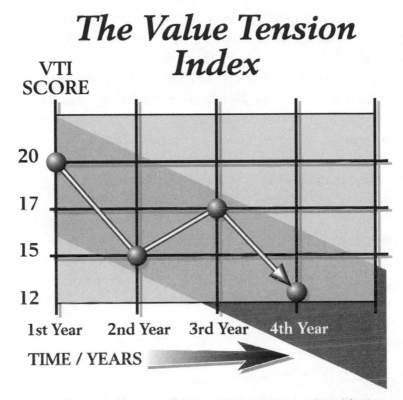

The Value Tension Index

The Value Tension Index typically improves (goes down) in the second year of a Shared Values effort. Many companies experience a reversal in the third year, the result of rising expectations by employees. Thereafter, improvement continues.

BOTTOM-LINE FINANCIAL BENEFITS

Executives sometimes ask, "Can a return on investment or profit outcome be predicted for Shared Values?" Shared Values benchmarking measures elements that directly equate to net profit improvements. The bottom-line financial benefits of Shared Values have been illustrated earlier—both in the Pepsi-Cola story at the beginning of this book, and in the description of the Mitzel restaurant chain's experiences.

Shared Values enabled the executives and employees of the

Pepsi bottler to handle a crisis in a way that not only averted possible disaster, but let the company survive with its reputation not only unscathed but enhanced.

The experience of the restaurant chain provided an early demonstration that has since been repeated with a great many other companies: the direct correlation between the Shared Values Process and improved profitability. Without exception, every company where Shared Values has been instituted has achieved not just a better work environment and more satisfied customers, but significantly greater financial success.

Two recent independent studies have reached the same conclusion. Both tracked a number of companies over a five-year period and found that "issues pertaining to trust, employee empowerment, participation, etc., lead to better overall financial performance."

WRAP UP

Benchmarking your work environment is a continual process. We recommend that you consider a new look at your organization every 14 to 18 months.

When you benchmark, be prepared to act. Performance in the workplace is linked with performance in the marketplace, and there may not be any better way of increasing cynicism among employees than conducting a benchmarking survey and then ignoring the results.

The Greek philosopher Plato suggested that wisdom begins when a person finds out that he does not know what he thought he knew. Benchmarking is the beginning of finding out.

And if you want to build enthusiasm for a Shared Values environment in your organization, gather data that will enable you to show your organization that Shared Values promises not just improvements in the workplace, but improvements in the bottom line as well.

1 8

Attaining Lasting
Change Through
Shared Values

Experience has ways of "boiling over" and making us correct our
present formulas.

Lloyd Morris

After ten years of introducing the Shared Values Process
into companies, we have come to recognize that Shared
Values is the most overlooked approach to running a
business and dealing with employees. Somehow, over the past
70 years, the concept of values has been overshadowed by scientific management, behaviorism, Darwinian economics, technology, strategy, expansions, mergers, acquisitions, improved
efficiency, productivity, total quality, human potential, goal setting, ethics, visioning, empowerment . . . and the list goes on.

In management books, the topic of values is often relegated
to a mention in the preface or brought up in connection with
some other topic. Why so little emphasis on such an important
subject? Even the University of Chicago group from which we

obtained our initial database failed to recognize the value of what it held. And yet, the idea of values, and of shared values, is becoming recognized as among the key ideas of our time—witness the success of William Bennett's *The Book of Virtues*, and Ambassador Richard Capen's *Finish Strong*. Presidential campaign speeches are replete with references to the subject, and it now crops up everywhere from the pages of *Fortune* and *Time* to the network evening news.

In the business community, Shared Values is finally becoming recognized by many as the key to the *lasting change* they have been searching for.

Company size has little to do with how Shared Values is introduced into an operation. What counts most is the commitment of the individuals involved. Though Shared Values has been recognized for its power, until now there has never been a prescription, a road map, for putting it into action in the workplace.

PROFIT FROM EXPERIENCE

The experience of instituting a Shared Values workplace at Century Circuits & Electronics, St. Paul, brought more than a few surprises, in the view of Tim Yungers, the company's manager of Human Resources. And in the surprises are lessons for other companies that will follow this path.

Prime example: the effort to create more equitable compensation plans, which proved an eye-opening experience for many of the employees. "In the past, this would have been researched and designed by HR, approved by management, and announced to the employees," Yungers says. "This time, with the task given to the employees themselves, they were surprised by the extent of the research and work it demanded." There are thousands of compensation systems, they discovered, and "each year some journal articles appear promoting new ideas that are supposed to be the latest, greatest and best."

The compensation team found out that "each individual has his or her own ideas of what would be the ideal system," Yungers says, and devising new plans "could be an explosive issue" demanding skills of listening, reasoning, and compromising.

But in some ways the process has been even harder for managers. "It demands a letting go. I used to be at a company where I helped develop all the systems. As a senior manager here, having to let go takes a leap of faith, and an understanding that the employees will do what's right not only for themselves, but for the company."

Besides providing a valuable and beneficial experience for the employees, and holding the promise of plans that all employees could support, the outcome brought unanticipated benefits. "Companies in high-tech are having a very difficult time recruiting people," Yungers observes. "You have to be able to sell not only an opportunity, but a company that works with its people and takes care of its people. And when you have the right People Systems, you've got something powerful to offer that other companies don't have."

Has the Shared Values effort at Century Circuits been worthwhile? "It's hard work," says Yungers, "but the strides we've made so far have been exceptional." (Work in process has been reduced by over 400 percent.)

KEEPING SHARED VALUES ALIVE

At the beginning of each new season, the great Green Bay Packers football coach Vince Lombardi would gather his players and begin, "Gentlemen, this is a football."

In the same way, establishing a Shared Values environment is not an end but a beginning—an opportunity for corporate leadership to reintroduce quality values at the workplace and know that these values will carry over into the family, the community, and the country.

When co-author Rob Lebow was a marketing manager at

Avon years ago, all employees took part in the company's annual golf tournament. Even those who didn't play golf joined in
on the social aspects of the day. The event drew the entire company together and provided an opportunity to reflect on the
traditions, heritage, and beliefs of the organization. This annual
reminder played a valuable role in keeping the heritage alive; in
times of crisis, Avon managers would remember to turn to the
written values of the company, asking themselves, "How can
these intrinsic values guide me now?"

In the same way, Apple Computer for years had a "beer
bash" every Friday; IBM has company outings. These events
have a staged informality, but serve an invaluable function of
keeping values and traditions fresh.

Reminders that keep Shared Values pulsating throughout a
company can be an intrinsic part of the informal occasions as
well as the formal meetings. The everyday events become landmark occasions when Shared Values messages are considered
as important as gross margin reports.

So the road to lasting change must continually be repaved
and kept in repair. You'll do this by:

- Adopting a set of values that your people have a hand in
  developing—a set of values they understand, accept,
  share, and apply.
- Maintaining standards that are non-negotiable and that
  provide guidance for professional actions, and personal
  actions as well. Ethics and integrity are not for advertising
  slogans, but are internal filters of our actions and
  behaviors.
- Providing regular ceremonies and channels—newsletters,
  communications meetings, picnics, parties, corporate
  videos, and so on—that sustain the established Shared
  Values.
- Recognizing individuals who exemplify the desired
  behaviors. Modeling behavior impacts the context of the

company and is therefore a key element in every successful Shared Values work environment. No pretense, just real people working together, trying hard to support each other.

- Monitoring the ongoing progress and successes by redoing the Values & Attitude Study at regular intervals. Measuring and checkpoints along the way are essential. Open interactions, without fingers pointed judgmentally, are vital if the atmosphere is to remain safe.

Finally, there are five commitments that will go a long way to ensuring the success of your Shared Values effort:

1. Make a personal commitment.

   This must be done by everyone throughout the organization, starting with the leader. There are no "experts." Remember, a Shared Values organization is a social experiment in real time.

2. Over-communicate the vision of Shared Values.

   A Shared Values environment needs to create a "pageantry" around the organization—just as each year we repeat the Thanksgiving dinner, the Christmas pageant, the Memorial Day parade, partly to teach the young, and partly to reconfirm what we believe.

3. Look for every possible *teachable moment.*

   Teachable moments are the most effective way for helping people understand the principles, and they are the fastest road to success. Make every teachable moment count. You can't remind people of values too often.

4. Root out elite groups.

   Elite groups are based on an "us versus them" notion, or, more often, "We are *better* than they are." Elite groups undermine the principles of Shared Values.

5. Accept that slow is better than fast: there are no time lines.

It doesn't do to say, "I wrote a memo on that last year; how come they don't get it?" Even with the best of intentions, a Shared Values environment doesn't arrive overnight.

Not long ago, a frustrated employee wrote a letter to the CEO of his company; in part the letter read as follows:

> After much thought and deliberation I found it necessary to write this letter which states my position. As you know I was promised, by you and the board of directors, an employment contract. You gave me a sample contract you indicated that I was to use as a starting point and go from there. Because I felt the model contract was generous, I in fact scaled it down. When I attempted to submit this contract to you, I was told the timing was not good. Rather than push the issue I decided to wait further developments.
>
> The recent activity with ___ now makes it imperative that we resolve this matter since the key clause of the contract is the termination of benefits due to me if ___ is acquired. I am convinced my contributions to the company were crucial to increasing the value of the company which now make[s] a profitable sale feasible.
>
> Your verbal discussions with me on this matter I consider to be a binding promise. I would have obtained it immediately in writing had you not been my father.

In a company where one staff member has difficulty being treated honorably by his own father, how do the rest of the employees fare? But it's not necessary to endure the chains and miseries of wretched working conditions any longer: the Shared Values environment promises a new, brighter, more fulfilling workplace.

WRAP UP

We end this book where we began: imagine a work environment where everyone puts the interests of others first, where managers and employees freely mentor one another, where taking risks is encouraged and people are given credit for their accomplishments. A place where everyone is open to new ideas, truth is common, and trust abounds.

Welcome to the Shared Values environment, a place of Lasting Change.

Index

AUTHOR CONTACT INFORMATION

For further information about the Shared Values Process in an organizational setting, distributor and associate opportunities to facilitate delivering shared values to your clients, benchmarking your work environment, or keynote speeches on the profound changes that shared values can bring to your organization, contact:

The Lebow Company

800-423-9327

425-828-3509

Fax: 425-828-3552

lebow@nwlink.com

lebow@aol.com

Or visit the Shared Values Web site at http://sharedvalues.com.